Elements of Farm Practice

by A.D. Wilson and E.W. Wilson

with an introduction by Roger Chambers

This work contains material that was originally published in 1915.

This publication was created and published for the public benefit, utilizing public funding and is within the Public Domain.

This edition is reprinted for educational purposes and in accordance with all applicable Federal Laws.

Introduction Copyright 2018 by Roger Chambers

Self Reliance Books

Get more historic titles on animal and stock breeding, gardening and old fashioned skills by visiting us at:

http://selfreliancebooks.blogspot.com/

Introduction

I am pleased to present yet another title in our "How To ..." series.

The work is in the Public Domain and is re-printed here in accordance with Federal Laws.

As with all reprinted books of this age that are intended to perfectly reproduce the original edition, considerable pains and effort had to be undertaken to correct fading and sometimes outright damage to existing proofs of this title. At times, this task is quite monumental, requiring an almost total "rebuilding" of some pages from digital proofs of multiple copies. Despite this, imperfections still sometimes exist in the final proof and may detract from the visual appearance of the text.

I hope you enjoy reading this book as much as I enjoyed making it available to readers again.

Roger Chambers

PREFACE

This book has been prepared primarily for use in rural schools and for elementary classes in other schools, wherever it is desirable to study the plain and practical problems of the farm and home in their relation to daily life.

The book deals largely with common farm practices, rather than with scientific principles. It is intended to throw some light upon and add interest to the things that are done on the farm from day to day. We feel that, if pupils can be interested and enabled to use the farms and the farm homes as laboratories in which to observe and apply the things learned at school, a great step will have been made toward bringing the school in close touch with the home life of the pupils.

A study of agriculture in the rural schools elevates, in the minds of the farm youth, the calling of agriculture. A rather close study of a few farm problems impresses one with the great amount of knowledge and skill required to operate a farm successfully; and must convince one that a farm, rightly managed, affords as much opportunity for development as a professional or business career.

We do not suggest that the topics shall necessarily be taken up in the order presented, but that each teacher begin with that portion of the book dealing with the particular phase of farm work that is being done on the home farms of the pupils at the time the lesson is given.

Each section is a complete reading lesson, followed by questions and examples, which relate to the lesson. The questions may be answered orally or in the form of a language paper. The examples will enforce some of the main facts taught in the text. This manner of presenting the various subjects was chosen so that the study of agriculture might replace a part of the regular reading, language and arithmetic lessons, and thus allow a study of agriculture without shortening the time of, or crowding out, any other subject.

Many of the complex problems encountered in the management of the farm are discussed here with a view to simplifying them, so that any one may easily understand the principles involved. It may even prove valuable to farm managers, by enabling them to put into practice some of the better methods of soil and live stock management, and to see clearly the aspects of farming as a business.

The idea of preparing these lessons originated with Mr. D. A. Wallace, editor of The Farmer, and we gratefully acknowledge his suggestions.

We have freely used many of the agricultural books and bulletins in the library at the Minnesota Agricultural College, and have obtained much valuable information therefrom.

Nearly all photographs used were made by Mr. H. D. Ayer, and the drawings were made by Mr. C. H. Welch and Mr. G. F. Krogh.

A. D. WILSON
E. W. WILSON

University Farm, St. Paul, Minn.,
    September, 1915.

CONTENTS

Chapter		Page
I	**Soils**	9
	Origin, Classification, Sources of Plant Food, Available Plant Food.	
II	**Tillage**	19
	Objects of Plowing, Time to Plow, Dry Farming, The Seed Bed, Planting.	
III	**Fertilizers**	31
	Need of Fertilizers, Fertilizers and Their Use, Animal Manures.	
IV	**Grain Crops**	39
	Plant Structure, Good Seed, Selection of Good Seed, Wheat, Oats, Barley.	
V	**Cultivated Crops**	52
	Corn, General Features of the Corn Crop, Shapes of Kernels of Corn, Sizes of Kernels of Corn, Parts of a Kernel of Corn, Testing Seed Corn for Germination, Corn Culture, Reasons for Cultivating Corn, Methods of Cultivating Corn, Selection of Seed Corn, How to Select Seed Corn, Storing Seed Corn, Methods of Storing Seed Corn, Corn for Silage, The Potato Crop, Planting and Cultivating Potatoes, Root Crops.	
VI	**Hay and Pasture Crops**	89
	Importance of Hay Crop, Clover, Clover Roots and Bacteria, Curing Hay, Alfalfa, Other Common Hay and Pasture Crops.	
VII	**Miscellaneous Crops**	108
	Millet, Rape, Field Peas, The Soy Bean, Cowpeas, Vetch, Rice, Sugar Cane, Fiber Crops, Cotton, Flax, Hemp.	
VIII	**Common Weeds and Their Eradication**	115
	Weeds, Weed Seeds Common in Grain, Weed Seeds Common in Grass and Clover, Classes of Weeds.	
IX	**The Garden**	125
	A Garden. Plan of a Garden. Some Common Vegetables.	
X	**Fruit on the Farm**	134
	Value of Fruit in the Diet, Strawberries, Raspberries, Apples.	

Chapter		Page
XI	**Plant Diseases and Insect Pests**............................	146
	Plant Diseases, Diseases of Potato and Cotton, Insects and Their Control.	
XII	Live Stock...	155
	Importance, Relation to the Soil, Classes, Care and Management.	
XIII	**Feeds and Feeding**..	163
	Source, Requirement, Selection, Kinds, Composition, Balanced Ration, Comparison of Feeds.	
XIV	**Horses**...	167
	Types and Breeds, Care and Management, Feeding.	
XV	**Cattle**...	182
	Types and Breeds, Care and Management, Feeding.	
XVI	**Dairying**...	202
	Milk and Its Care, Testing Milk, Testing Cows.	
XVII	**Sheep**...	210
	Types and Breeds, Care and Management, Feeding.	
XVIII	**Swine**...	218
	Types and Breeds, Care and Management, Feeding.	
XIX	**Poultry, Birds and Bees**.................................	232
	Poultry on the Farm. Care of Poultry, A 100-Hen Poultry House, Feeding Laying Hens, Birds, Bees.	
XX	**Agricultural Engineering**................................	247
	The Road Problem, Road Construction, Maintenance of Roads, Drainage, Irrigation, Farm Machinery, Farm Buildings, The Silo, Fencing, Building Fences.	
XXI	**Community Activities**...................................	272
	Boys' and Girls' Clubs, Farmers' Clubs, Co-operation, Marketing Butter, Marketing Eggs, School Gardens, County Agent.	
XXII	**The Farm Home**...	293
	What a Desirable Home Should Be, Windbreaks, Sanitation, Ventilation, The Farmstead.	
XXIII	**Farm Management**.......................................	306
	Standing of the Farmer, Rotation of Crops, Classification of Field Crops, Rotation Maintains Vegetable Matter, Planning Farms, Live Stock Accounts, Account with a Cow, Marketing Dairy Products, Co-operation in Delivering Milk or Cream.	

ELEMENTS OF FARM PRACTICE

CHAPTER I

SOILS

Soil, from the standpoint of the farmer is that portion of the earth's surface in which plants grow. It is composed of small particles of rock, as grains of clay and sand, and decayed and decaying plants.

Origin.—We are told that at one time, many, many years ago, the earth's surface was all solid rock, and that the wind and water and frost have been able to break off little pieces of rock to make soil. These little pieces of rock are called clay when very fine, sand when a little coarser than clay, and gravel when quite coarse. In the mountains, or where there are very large stones or boulders,

Figure 1.—The action of vegetation, water and the weather gradually causes the disintegration of solid rock.

large cracks will be seen in the rocks. These cracks are made by frost, by alternate expansion and contraction caused by heat and cooling, or by the force of growing roots. When these cracks are formed little particles of rock are broken off. A strong wind will blow these particles about over the rocks, make them finer and wear off other particles. Rain and water running over the rocks do the same thing. The wind and water tend to gather the soil particles into crevices in the rock and other sheltered places. When several of these little particles

Figure 2.—An excellent specimen of humus.

have been gathered in one place, there is the beginning of a little patch of soil. When this little patch of soil becomes moist from rain or melting snow, and the warm sun shines on it, some kind of a plant, like moss, will start to grow. At first these little patches are very small and plants can grow only a very little while. When the plants die they are added to the soil. Then other small pieces of rock are added and still other plants grow and die and are added to the little patch of soil. This has been going on for many thousands of years, so that nearly the whole surface of the earth is covered with soil.

Parts of Soil.—All soils are then composed of two parts, the part made up of little particles of rock that we call sand or clay or gravel, and the part made up of decayed or decaying plants. This part is called organic matter, vegetable matter or humus. Wherever crops are to grow it is necessary that the soil have both particles of rock and vegetable matter. In a sand pit there is no vegetable matter in the soil and plants grow very poorly or not at all. In an old drained lake-bed, where the soil is made up almost entirely of vegetable matter (peat), crops do not grow well. In farming it is very important that there be a proper combination of these two parts. The part made up of particles of rock is called mineral matter. The part made up of dead and partly decayed plants is called organic matter. Plants contain both mineral matter and organic matter. If dried plants are burned, the part that passes off in the air is the organic matter. The part that is left, as ashes, is mineral matter.

Questions:
1. What is soil?
2. How has soil gradually been formed from solid rock?
3. What are the two important parts of soil?

Arithmetic:
1. If there are 2 lbs. of ash in 100 lbs. of dry vegetable matter, how many lbs. of ash in one ton (2,000 lbs.) of vegetable matter?
2. If there is ½ lb. of vegetable matter in 10 lbs. of soil, how many pounds of vegetable matter in 100 lbs. of soil?

CLASSIFICATION OF SOILS

In the study of soils one finds that there are many different kinds. To enable us to talk and write about soils, and to understand what is meant, it is necessary to classify soils so that everyone will know what is meant when a certain kind of soil is named. The most common names applied to soils are gravel, sand, loam, clay and peat.

Gravel is the coarse part of the soil. The particles may vary in size from that of kernels of wheat to stones as large as hen's eggs. Such soil is not as a rule productive. The particles are so coarse that they hold very little moisture or plant food. Soil containing a large percentage of gravel is called very poor or very light.

Sand is the name applied to soil with particles much finer than gravel but still comparatively coarse. The particles may be as large as grains of common granulated sugar or cornmeal. Sandy soil is much more productive than gravelly soils. The particles being finer, they hold moisture better and usually contain more available plant food.

Clay is the name applied to the very fine particles of soil. Clay is often as fine or finer than wheat flour. Generally clay soils are the most productive soils, because the grains are very, very fine. A given quantity of clay will hold much more water than the same quantity of sand or gravel. Clay soil is very sticky when wet, while sandy soil is not.

Loam is a name applied to soil that has a liberal amount of vegetable matter mixed with either sand or clay or both. If a soil has a very large proportion of clay, it is called a clayey loam. Nearly all the soils have a mixture of sand with clay, but the proportion of each naturally varies. On this account there are all kinds of mixtures, varying from nearly all sand and very little clay, to nearly all clay and very little sand.

Sandy Soils.—Soils containing a large percentage of sand are known as sandy soils, or sandy loam soils. They do not hold as much moisture as clay soils. Such soils, therefore, warm up more quickly in the spring than do clay soils, and crops grow more quickly. If it does not rain for several days or weeks, crops on such soils are likely to be injured for lack of moisture. Sandy soils contain less plant food than clay soils and give it up more readily. On this account clay soils are regarded as better; but, if sandy soils are well handled, they produce good crops and are more easily plowed and cared for than clay soils.

Clay Soils.—Soils containing a large percentage of clay are known as clay soils, or as clay loam soils. Because the particles of clay are very small, there is more surface exposed in a given amount of clay soil than in the same amount of sandy soil. Soil holds water on the surface of the particles, on which account a clay soil holds much more water than a sandy soil. The fact that a soil with fine particles has more surface exposed than soil with coarser particles is illustrated by an apple. The surface of a whole apple is represented by the peeling. If the apple is quartered, or cut into many pieces, each cut increases the exposed surface of the apple by the two newly cut surfaces. The apple is no larger. The exposed surface represented by the peeling is the same. If a grain of sand is pulverized to form many particles of clay, the amount of surface will be greatly increased. Because clay holds moisture better, it warms up more slowly in the spring. Crops start more slowly, but are much less likely to be injured by drouth.

Questions:
1. What do you understand by the terms, gravel, sand, clay?
2. Tell the difference between a sandy loam, and a clay loam.
3. Explain why clay soil holds moisture better than sandy soil.

Arithmetic:
1. If a cubic foot of sandy soil weighs 90 lbs. and holds 17% of its weight of water, how many pounds of water will it hold?
2. If a cubic foot of clay soil weighs 75 lbs. and holds 30% of its weight of water, how many pounds of water will it hold?

SOURCES OF PLANT FOOD

Plant Food in the Air.—Plants as well as animals must have food; and it is as important to know what plants need and how to supply their needs as it is to know how to feed animals properly.

The greater portion of the plant food comes from the air rather than from the soil. All those substances in a plant called carbohydrates, as starch, sugar and fibrous tissue, are made entirely from carbon dioxide gas and water. The plant takes in carbon dioxide from the air, through its leaves, and water from the soil, through its roots. When the water and the carbon dioxide are brought together in the leaves of the plant, and the sun shines on the leaves, the sun and the green coloring matter (the chlorophyll) in the leaves cause the water and the carbon dioxide to unite. The oxygen and hydrogen in the water unite with the carbon in the carbon dioxide. These three elements form starch. The oxygen in the carbon dioxide is liberated and given off to the air. In this way plants purify the air for animals to breathe and animals exhale air containing carbon dioxide, which furnishes food for plants. Some of the starch formed in a plant is slightly modified during the growth of the plant and forms fibrous tissue and sugar. Examine kernels of wheat and corn and a potato to see what a very large part is starch. The white part of them all is very largely starch. It is seen that by far the greater portion of our common plants does not come from the soil, as is usually supposed, but is formed from the poisonous gas, carbon dioxide, from the air, and water from the soil.

Plant Food in the Soil.—A small portion of every plant comes from the plant food in the soil. A fairly good idea of the proportion of any plant that is taken from the soil is obtained by burning the plant. The ashes remaining represent nearly the whole amount that came from the soil. This portion, though small, is absolutely necessary for plant growth. One may liken the plant food taken from the soil to salt eaten by animals.

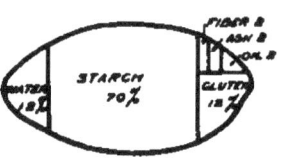

Figure 3.—The approximate composition of wheat. The ash and a small part of the gluten are composed of minerals taken from the soil. The rest is composed of carbon dioxide from the air, and water.

It furnishes a very small part of the food required, but is absolutely necessary. Hence the importance of having a fertile soil that will furnish these substances as needed by the growing crop.

Plants are able to get food from the soil only when it is in a soluble form—that is, when the plant food will dissolve in water as sugar dissolves in tea.

Soluble Plant Food.—When a soil contains plenty of soluble plant food it is said to be fertile. When plant food becomes soluble in the soil it is dissolved in the soil water. This water containing the plant food, surrounds the fine roots and root hairs of the plant, and passes through their thin walls just as nourishment enters the blood vessels in the animal body. In this way plants get their food, soluble organic matter and soluble mineral matter, from the soil. They take in the plant food with

Figure 4.—A diagram showing the composition of a potato. From Minnesota Bulletin No. 42.

large quantities of water. The water is given off from the leaves into the air, leaving the plant food taken from the soil in the plant. To show plainly that liquid passes up through the stem of a plant and into the leaves, set a branch from a house plant into a bottle of red ink or colored liquid, and watch results. It is seen that large amounts of water are needed by growing plants. Scientists have shown that to produce one pound of dry product, as hay or corn fodder, a plant takes from the soil and gives off to the air from 200 to 500 pounds of water.

Since plants use water from which to make starch and other similar substances, as well as large quantities to take up the other plant food, it is very necessary that they be supplied with sufficient water at all times.

Soil Moisture.—There is in most places enough moisture from rainfall or irrigation to produce good crops, but it is not always available at the right time; and often, during the growing season, there may be a shortage for several weeks. Unless land is in good condition to hold moisture, and well cultivated to prevent evaporation from the surface, it may become too dry, and then the plants will not grow well or may die. Farmers can avoid this difficulty largely by keeping vegetable matter in the soil, which holds moisture like a sponge; and by thorough cultivation of the surface, which prevents, to a large extent, the loss of soil water by evaporation. By cultivation the soil is loosened at the surface and the water in the ground cannot rise readily by capillarity, because it is separated from the sun and wind by this layer of loose soil.

It is in such times that the skillful farmer, or the man who knows best how to handle his soil, can get good crops, when farmers who do not know or care, but just "trust to luck," fail.

Questions:
1. Of what substances are plants largely composed?
2. Tell how these substances are converted into plant tissue.
3. From what source does a plant get a small but essential portion of its food?

Arithmetic:
1. How many pounds of wheat are produced on an acre yielding 20 bus.? (A bu. of wheat weighs 60 lbs.)

16   ELEMENTS OF FARM PRACTICE

2. How many pounds of corn are produced on an acre yielding 40 bus.? (A bu. of shelled corn weighs 56 lbs.)

3. How many pounds of potatoes are produced on an acre yielding 150 bus.? (A bu. of potatoes weighs 60 lbs.)

AVAILABLE PLANT FOOD

Amount of Plant Food.—Most soils contain enough plant food to grow crops for many years, several hundred perhaps, but this plant food is not present in the soil in a soluble form and it is well it is not. If it were soluble it would be dissolved by the water during a heavy rain, and as the water flowed off over the fields and into the river it would carry with it the plant food, thus leaving the soil unproductive. This may be better understood if one takes two glasses of water, puts a spoonful of sand into one and a spoonful of sugar into the other, and stirs. Then carefully pour the water out of both glasses. The sugar being soluble has been dissolved and will pass out

Figure 5.—(1) A piece of sod showing the grass roots. (2) A lump of earth taken from a field that has grown corn continuously for fourteen years and which is badly depleted of vegetable matter.

of the glass with the water. The sand is not soluble and will remain in the glass.

Only a very small amount of soluble plant food is needed to grow a crop; but while the amount is small, it is absolutely necessary to have enough of it to supply the plants.

How Plant Food Is Made Soluble.—There are many different ways of making the insoluble plant food in the soil soluble. These are Nature's ways, and the change takes place naturally in soils under favorable conditions. But farmers can do a great many things to assist Nature in this work.

One very important condition of soil, which favors making plant food soluble, is to keep the soil well supplied with vegetable matter as it was when the farmer first broke up the virgin sod. Get a small piece of sod from a new piece of breaking, and a handful of soil from an old field that has grown nothing but corn or grain for a great many years. Notice that the first is tough and is held together by many fine roots interwoven among the soil grains. The handful of earth from the old field contains little except the particles of soil.

The plant roots, as well as other parts of plants found in soil, are called vegetable matter. When this vegetable matter is partly decomposed, it is called humus.

Decay of Vegetable Matter.—When the weather is warm and the soil moist, the vegetable matter in the soil begins to decay. The vegetable matter is composed of plants, and is made up of the things that growing plants need for food. When the vegetable matter decays, the substances of which it is composed are set free or liberated, thus making plant food soluble. The vegetable matter decaying in the soil not only liberates the plant food of which it is composed, but aids very much in making some of the insoluble plant foods in the mineral particles of soil soluble. It also aids by making the soil warmer, as heat is given off by a decomposing manure pile, and by giving off an acid, called organic acid, because it is formed from organic matter. This acid acts on the soil grains and dissolves a small amount of mineral matter off their surfaces.

Plant food is thus made soluble much more rapidly in a soil that contains a good supply of vegetable matter, as new sod land, than in an old soil from which much of the vegetable matter has been used.

Many soils that have produced grain and corn a great many years without the addition of manure have become nearly depleted of vegetable matter; and, while they usually contain plenty of plant food, it is in an insoluble form and plants cannot make use of it.

Adding Vegetable Matter.—A farmer can add vegetable matter to his soil by growing on it once in every few years such crops as clover, timothy and other grass crops. These crops grow more than one year and consequently have a large root system. These roots add a large supply of vegetable matter; so several crops of corn or grain can be grown successfully following a crop of grass. The application of barnyard manure is another way by which the farmer can put vegetable matter in the soil and thereby increase its producing power.

Examine carefully a newly plowed field that has recently grown a crop of tame grass. If possible note the growth of the crop, from time to time, on such a field, and compare it with the same kind of a crop on an old piece of land that has not grown a grass crop or been manured for several years.

Questions:
1. What do you understand by the term vegetable matter in the soil?
2. In what way does vegetable matter assist in making plant food soluble?
3. In what ways may a farmer add vegetable matter to the soil?

Arithmetic:
1. How many lbs. of water in 20 bus. of wheat?
(Note: There are 12 lbs. of water in 100 lbs. of wheat.)
2. If 70% of wheat is starch, how many lbs. of starch in 20 bus.?
3. It requires 500 lbs. of water to produce one pound of hay. How much water is required to produce a ton of hay?

CHAPTER II
TILLAGE
OBJECTS OF PLOWING

Plowing Not Monotonous.—Did you ever wonder as you watched men plowing, why they were doing it? Or did you ever think that plowing must be very monotonous work—going forth and back, forth and back, across the field day after day? Plowing is not unpleasant work. In fact, most men like to plow. It is a quiet, peaceful work, and after the rush and anxiety of harvest time, it really seems restful. It is certainly not monotonous work, if one knows why one is plowing, and how and when to plow.

Plowing Mellows the Soil.—One of the chief reasons for plowing is to stir the soil and make it loose and mellow, so the air can circulate through it, and so the moisture can settle down into it. When the soil has settled all summer and had the heavy rains beating on it, it becomes

Figure 6. Good plowing done with a breaker.

packed and hard, and must be loosened to prepare it for a new crop, if we desire to meet the conditions of nature.

Plowing saves moisture for the next crop. When the ground is packed, as it is when the crops are removed in the fall, it is so hard that when it rains much of the water runs off over the surface instead of settling down into the soil. Plowing overcomes this difficulty, and by loosening the surface, any moisture that may be in the subsoil (the soil below the plowed furrow) is retained, because this water cannot readily pass up through the loose plowed soil. If the ground were not plowed, the soil moisture would rise to the surface by capillarity, just as oil rises in a lampwick, and when it got near the surface the sun and wind would evaporate it.

Plowing Destroys Weeds.—It is natural for all good soils to be producing something at all times during the growing season. As soon as the crop is removed (and very often before) weeds begin to grow. If no precaution is taken, they will go to seed, and thus cause trouble later. Plowing stops their growth. It also turns up new soil to the light, and weed seeds which have been too deep in the soil to grow, are brought near the surface where they can grow. If the plowing is done in the fall these weeds start to grow, but do not have time to produce seed before they are killed by frost.

Figure 7.—First year clover growing in stubble. Such a crop is better pasture than many cattle are furnished, and is a benefit to the soil. It would be unwise to plow such a field early in the fall, if one has stock that can use the feed.

Plowing Destroys Insects.—Many insects, such as grass hoppers and cutworms, are checked by plowing in the fall. The mature insects lay their eggs in the ground in the fall, and if the eggs are not disturbed, they hatch out the next spring, and it is the young from these eggs that do the damage. Fall plowing disturbs the eggs and many of them are destroyed.

Plowing Helps to Liberate Plant Food.—We learned in another lesson that plants require food in a soluble form; that is, food in such condition that it will dissolve in water. Plowing assists in making portions of the soil soluble, by pulverizing it, breaking up the soil particles, exposing new surfaces, and allowing the sun, wind and water to act on it more freely than they can act on unplowed land.

Plowing Covers Manure and Crop Residue.—It is generally regarded as good practice to haul manure directly from the barn to the field before it has rotted, as much of its value is saved in this way. If a rather heavy dressing is applied, it is troublesome in harrowing, sowing and cultivating, unless it is plowed under out of reach of the harrow and other tools, but still where the plant roots can reach it.

It will be interesting to note the effect of plowing, by observing the work as it is being done in some conveniently located field.

Questions:
1. Explain how plowing saves moisture.
2. In what two ways does plowing destroy weeds?
3. How does plowing destroy insects?
4. How does plowing assist in liberating plant food?

Arithmetic:
1. A plow turns a furrow 14 inches wide. How many furrows must one plow to plow a strip 8 rods wide?
2. How far will a team travel in plowing with a single 14-in. plow a field 8 rods wide and 40 rods long?
3. How many acres of land in a field 8 rods by 40 rods?

TIME TO PLOW

Condition of Soil.—The greatest problem that a plowman has to solve, is to determine the proper time to plow. Both the season of the year and the condition of the soil must be considered. If a heavy, clayey soil is plowed when it is too wet, the lumps turned up become, when dry, hard clods, which it is very difficult to pulverize into a good seed bed. This is especially true if plowing is done in the spring. Plowing clayey land that is wet is not objectionable, however, if done in the fall and the field is not sown until spring, as the thawing and freezing during winter aid in pulverizing any clods that may form. Light sandy or loam soil may be plowed when wet without any serious trouble.

Time of Year to Plow.—As a rule early fall plowing is preferable, as it allows the ground to become settled before the crop is sown, thus making it less likely to become too dry during the summer. Early fall plowing also destroys weeds by turning up new seeds, which start to grow in the

Figure 8.—Rape growing in a stubble field. Such a crop may be raised for fall feed at a very small cost per acre. Rape is excellent feed for any kind of stock but milch cows. One might be justified in neglecting to plow such a field early.

fall, and are soon killed by frost, while, if the same seeds were turned up late in the fall, the plants would grow in the spring and trouble the crops. Fall plowing also facilitates spring work, and, by leaving the soil exposed to the elements, aids in liberating plant food.

When Not to Plow in Early Fall.—If some catch crop is growing in the stubble field, as clover, rape or rye, that can be used to advantage for fall pasture, fall plowing—at least early fall plowing—is not always advisable. The green crop and pasturing will prevent largely the growth of weeds, and the green crop checks to some extent the loss of moisture. Thus at least two reasons for early plowing are removed. The green crop, whether plowed under or pastured off, would add considerable vegetable matter to the soil, which would be of more value to the next crop than the plant food that would be liberated by the early fall plowing. As the country becomes more thickly settled, and better methods of farming are practiced, farmers will have fewer fields lying idle during the fall. Many farmers now get from 50c to $4.00 worth of feed per acre from their fields after the main crop has been harvested. This income is almost entirely net profit, and cannot be overlooked as more intensive systems of farming are made necessary by higher priced land.

Fall Feed.—On many farms pastures are very poor during the fall and cattle must be fed dry feed or, what more often follows, allowed to get poor or to run down in milk flow. Such conditions are very undesirable, and in most years unnecessary. Clover sown with the grain crop in the spring, or rye sown in the stubble as soon as the grain crop is removed, or rape sown at almost any season of the year, will in ordinary years furnish an abundance of fall pasture. Good fall pasture not only furnishes cheap feed during the fall, but gives stock an excellent start for winter. While the old habit of getting all the land plowed in the fall was an excellent one, and necessary when grain was the only crop, there are now many instances where much better results would be obtained were some of the fields made to produce fall pasture rather than left bare during the fall, in which condition more or less plant food is lost by such exposure. If catch crops are grown on the field so that early fall plowing can not be done, it is better to plow late in the fall than to wait until spring. Most crops do better on fall plowing than on spring plowing. Spring plowing, being loose, is likely to become too dry.

No Best Time.—There is no best time to plow. The time must be determined by conditions. It is hoped that those who read this lesson will think about the things mentioned, and observe what the best farmers in their neighborhoods are doing. They will then be better able to decide intelligently when to plow.

Questions:
1. What danger is there in plowing a heavy clay soil when wet?
2. What is to be gained by early fall plowing?
3. Give some reasons which may make it advisable to defer plowing until late in the fall.

Arithmetic:
1. 4 lbs. of clover seed per acre is sufficient to sow with a grain crop for fall pasture. What is the cost per acre of such pasture if clover seed is worth 15c. per pound?
2. If one acre of such pasture furnishes feed for a cow for 20 days, the cow giving ¾ pounds of butter fat per day, how much butter fat is produced per acre? What is it worth at 28c. per pound?
3. Three pounds of rape seed per acre is sufficient to sow with a grain crop for fall pasture. What is the cost per acre of such a crop if rape seed is worth 18c. per pound?
4. One acre of such rape will feed 10 sheep for 1 week. They will gain 2 lbs. per week each. How many pounds of mutton are produced per acre? What is it worth at 7c. per pound?

DRY FARMING

Dry farming is a term applied to the culture of land where the rainfall is not sufficient to grow crops in the ordinary way. It consists in deep plowing, packing the lower part of the furrow slice, and the maintenance of a surface mulch by persistent cultivation of the surface soil to prevent evaporation. In many sections in the western and central western part of the United States, the rainfall is less than 20 inches per year. Ordinary methods of farming have not proved profitable in such places, because the moisture was not sufficient to ensure a crop. The aim in dry farming is to so handle the soil that every bit of moisture that falls will be taken up by the soil, and the loss of moisture by evaporation will be as small as possible. In some sections where the rainfall is fairly plentiful,—that is, from fifteen to twenty inches per year,—a crop is grown every year. In other places, where the rainfall is only ten to fifteen inches, a crop is grown every other year.

Deep plowing, or plowing from eight to twelve inches deep, provides a loose mellow soil into which any moisture that falls quickly settles. It provides also much more room for the storage of moisture than is provided by shallow plowing, and allows the deep rooting of crops grown so they can better get the moisture stored in the soil.

TILLAGE

Subsurface packing is another important feature of dry farming. It is done by means of specially made heavy implements that are drawn over the field and pack the lower part of the furrow slice, but leave the surface mellow and loose. This packing increases the capacity of the soil to hold moisture, and packs the furrow slice against the subsoil so that any moisture in the subsoil may be brought up into the furrow slice by capillarity.

A surface mulch consists of from two to four inches of loose, fine soil over the surface of the field. It is main-

Figure 9. A subsurface packer.

tained by persistent disking and harrowing. Every rain packs this surface soil down, and, if left packed down, the moisture in the lower part of the furrow slice or in the subsoil will be raised to the surface by capillarity. If brought to the surface the sun and wind will evaporate it, and it will be lost. Harrowing at once after a rain loosens up this surface soil, making of it a surface mulch through which the moisture cannot rise.

Experiment.—A very good illustration of the effect of a surface mulch in preventing the rise of moisture to the surface may be seen by taking a loaf of sugar, putting on top of it some pulverized sugar (not granulated), then set the loaf of sugar in shallow water, colored with ink, and watch the liquid rise up to the pulverized sugar, then stop. Why does the water not rise through the mulch?

Questions:
1. What do you understand by the term, "dry farming"?
2. What is the reason for deep plowing in dry farming?
3. What do you mean by the term, "surface mulch"? Of what use is it? How is it maintained?

Arithmetic:
1. How many gallons of water fall on an acre when there is a rainfall of 1 inch? (Note: There are 43,560 sq. ft. of surface in an acre. There are 231 cu. in. in a gallon.)
2. If a cubic foot of soil will hold 3 gallons of water by capillarity, how many gallons will an acre of soil to a depth of 2 feet hold? How many inches of rainfall would this represent?

THE SEED BED

Yields.—The way in which the seed bed is prepared has much to do with the success of the crop grown. Where farms are large there is a strong tendency to rush through the spring work and get in a large acreage of crops, but often without due preparation of the seed bed. Such hasty work at seeding time is very often the chief cause of a poor harvest.

A yield of twenty-five bushels of oats per acre leaves the farmer no profit, as it costs as much to raise them as they are worth. A yield of forty bushels per acre leaves considerable margin for profit. Twenty-five acres of oats yielding forty bushels per acre are much more profitable than forty acres yielding twenty-five bushels per acre, as the same amount of oats is raised with less land and labor.

A good seed bed must be moist (not wet), firm enough so that it will not dry out quickly, loose enough to permit air to enter the soil, and warm enough to cause the seed to germinate. The farmer cannot regulate the weather, but he can do many things to assist in regulating these conditions, and such is the object of tillage.

Air is needed in the soil to start the seed to germinate and to supply the oxygen necessary in the chemical action which must take place in the soil, to make the plant food in the seed available for the growing plantlet and to break down plant food in the soil on which the plantlet can feed after it has used up the food stored in the seed. Cultivation with a disk or harrow stirs up the soil and lets the air circulate through it.

TILLAGE

Moisture is needed in the soil—(1) to dissolve the plant food in the seed planted, so that the little plantlet can make use of it; (2) to supply the growing plant with water; (3) to assist in the chemical action in the soil which liberates plant food; (4) to carry the plant food to the plant. Cultivation of the soil helps to retain moisture by checking evaporation from the surface by means of the surface mulch and by loosening up the surface soil so that any rain that falls will settle into it instead of running off over the surface.

Figure 10.—Preparing the seed bed by disking.

Need of Heat.—Seed will not germinate, neither will plants grow, unless the soil has a certain amount of heat in it. Heat is necessary before chemical action can begin. One can not make the weather warmer, but cultivation of the soil, keeping it loose on top so as to check evaporation from the surface, helps to warm up the soil. The circulation of air, promoted by good tillage, is also quite a factor in warming the soil in the spring. The air at the surface of the ground becomes warmed by the sun, and if it can enter the soil it helps to warm that also.

Methods of Preparing the Seed Bed.—The best seed bed is formed by plowing land in the fall, so that the portion turned over by the plow will have a chance to settle down upon the soil beneath (the subsoil). Then moisture, which is usually present in the subsoil, may move up into the furrow slice by capillary action, as oil rises in a lampwick. This moisture is often necessary to supply growing crops during times when it does not rain for several days. Fall plowing, disked and harrowed to loosen the surface, makes an excellent seed bed. In other words, a firm, mellow soil below, covered by two or three inches of loose, fine soil is the condition desired.

If land must be plowed in the spring, very thorough harrowing is necessary to work the soil up fine and to assist in firming the furrow slice so as to form good capillary connection with the subsoil.

Close observation of fields in the process of preparation for seeding will illustrate forcefully the points mentioned above.

Questions:
1. What is the principal work of the farmer during April?
2. What are the essential conditions of a good seed bed?
3. Why is air needed in the soil, and how may it be secured?
4. Why is moisture needed in soil? How may it be retained?

Arithmetic:
1. If wheat is worth 90c. per bu. and it costs 15c. per acre to harrow land, how many times can one afford to harrow an acre of land to increase the yield two bushels?
2. If wheat is worth 90c. per bu. and it costs 35c. per acre to disk the land, how many times can one afford to disk an acre of land to increase the yield 2 bus. per acre?
3. Field A yields 25 bus. of oats, field B yields 40 bus. of oats. How many more dollars' worth of labor can one afford to put on field A than on field B, if oats are worth 35c. per bushel?

PLANTING

Time to Plant.—Crops that are not easily killed by frost, as wheat and oats, are usually the first crops sown. Seeds of these crops will germinate at a comparatively low temperature, as low as from 41 to 50 degrees F. The soil usually reaches this temperature in the spring about as soon as one can begin disking and harrowing, and land that is well disked and harrowed reaches this temperature earlier, as shown in the preceding lesson. It is usually wise to sow these crops as early as possible and thus avoid the danger from rust, smut and hot winds that are more likely to injure late sown grain crops.

Barley may be sown early, but it is more liable to injury from frost. Experiments show that the best yields are obtained by sowing barley a week or ten days later than the first seeding of wheat or oats. This is also the most convenient time, as it permits one to sow the other grains first, and then to prepare the barley land. Barley may be sown as late as the last of May, if necessary, in the case of low, wet land; but earlier sowing is better.

Figure 11.—Seeding with a drill. The dragging chains cover the seeds, which are placed in the ground at a uniform depth.

Depth to Plant.—There are two ways to sow grain. First, by a broadcast seeder, which scatters the seed on top of the ground, where it is covered by cultivating or by harrowing. In this process some seeds are left on the surface uncovered, while others are covered as deep as the land is cultivated. This causes the seeds to germinate unevenly; and, if the land becomes too dry, much of the seed on the surface will not grow, while, if the soil is too wet, much of the deeply sown seed will not grow. The better method of sowing is with a drill which deposits all the seed at a uniform depth and at any depth desired. In early seeding one should plant quite shallow, from one to two inches, as the soil is warmer near the surface and the seed and small plants should have all the heat available. On the other hand, seed sown too shallow will not grow well if the weather remains dry for some time, as the surface of the soil dries out too quickly and leaves the plant improperly supplied with moisture. Later in the season when the soil is warmer and plants grow more quickly, it is well to plant the seed deeper, from two to three inches, thus giving the plants a better chance to get moisture.

The depth of planting should vary also with the kind of soil. In light, dry soil one should sow deeper than in heavy, wet soil.

Experiment.—It is interesting to plant short rows of seeds at different depths and at different times, noting results. Early in spring plant four short rows of oats or wheat. In the first row, plant seeds one inch deep; in the second row, two inches deep; in the third row, three inches deep; and in the fourth row, four inches deep.

Plant an equal number of seeds in each row and note the time required for plants to come up, number of plants that grow and strength of plants. Repeat the experiment later when the soil is warmer.

Questions:
1. What three conditions must a seed have before it can grow?
2. What can you say of time of planting?
3. What are the advantages and disadvantages of shallow and deep planting?

Arithmetic:
1. If the average yield of wheat in the United States is 14.8 bus. per acre, what is the average value of an acre of wheat at 85c. per bushel?
2. If it costs $13.00 to produce an acre of wheat, what is the average profit per acre? (See example No. 1.)
3. If the average yield of oats in the United States is 30 bus. per acre, what is the value of an average acre of oats at 38c. per bu?
4. If it costs $13.00 to produce an acre of oats. what is the average profit per acre?

CHAPTER III
FERTILIZERS

Need of Fertilizers.—If a soil is cropped year after year and no plant food is added, the supply in the soil will become exhausted, or there will be such a small amount left that the crops cannot get enough food to grow well and will produce very little or fail entirely. There is a number of different materials that may be applied to the soil to supply the plant food needed. Such materials are called fertilizers. Barnyard manure is the most common fertilizer used. Materials that are purchased for fertilizer,

Figure 12.—Pasturing. Live stock can be kept ordinarily more cheaply on pastures than in any other way. Pasturing also improves the soil.

such as lime, ground phosphate rock, sodium nitrate, waste from slaughter houses, etc., are called commercial fertilizers, because they are bought and sold.

Plant Food.—Plants, like animals, have to be fed, and, like animals, they need more than one kind of food. There is a number of different elements which are needed for the complete growth of plants. The soil, air, and water in the soil furnish most of the elements needed in such great abundance that there is no danger of the supply's ever becoming exhausted. There are only four elements that are likely to be lacking even in soils that have been cropped for many centuries. So it is necessary to learn only about the ones that are most likely to be required.

Names.—The four elements that are most likely to become limited in the soil are nitrogen, phosphorus, potassium and calcium. The names of these elements are not hard to pronounce or learn. Every boy and girl should know these names, what part each element plays in the growth of plants, and a practical means of maintaining

Figure 13.—A crop of alfalfa, one of the best farm fertilizers.

a supply of them in the soil. Boys and girls who have studied physiology and have learned how to pronounce and to know the meaning of such words as occipital, parietal, cerebellum, etc., or who have learned to pronounce and know the meaning of such words as subtrahend and minuend in arithmetic will not have difficulty in learning to use the words, nitrogen, phosphorus, potassium and calcium.

Nitrogen is needed by all plants. Nearly 80% of the air is nitrogen, and this is the chief source of nitrogen for the soil. The problem of the farmer is to get the nitrogen out of the air and into the soil. It is present in the soil chiefly in combination with other elements in the form of vegetable matter. As soon as the vegetable matter in the soil decomposes (rots), the nitrogen is made soluble or goes off into the air as gas, so that it is very easily lost. A soil that is lacking in vegetable matter is likely to be lacking in nitrogen. A good supply of nitrogen in the soil stimulates the growth of plants. A dark green color of the growing plants generally indicates plenty of nitrogen. When the foliage turns yellow before it is ripe, there is usually a scarcity of nitrogen. Nitrogen is an essential element of protein. The amount in normal soils varies from 2,000 lbs. to 10,000 lbs. in the upper seven inches.

Phosphorus is needed by all plants. It is needed especially in crops like grain and corn that mature seeds. The presence of plenty of phosphorus in the soil aids plants in the production of seed. The kernels of wheat or rye or corn are likely to be plumper and heavier where there is a good supply of phosphorus. A lack of sufficient phosphorus in the soil results in a smaller yield of grain, and the grain produced is not so good in quality. Selling seeds, such as wheat, barley, rye, or corn from the farm removes from the farm comparatively large amounts of phosphorus. The amount of phosphorus in normal soils varies from 1,000 lbs. to 2,000 lbs. per acre. When one considers the comparatively small amount of this element in the soil, and the amounts removed by ordinary crops, as shown in the following table, one is impressed with the need of adding to the supply in the soil before it gets too low.

Approximate Amount of Fertility Removed by Crops.

Crop	Amount bushels	Nitrogen pounds	Phosphorus pounds	Potassium pounds
Wheat	20	22	4.1	5.8
Oats	40	28	4.1	6.6
Barley	30	27	5.2	6.1
Corn	50	31	6.1	9.6
Potatoes	150	32	6.3	12.0
Clover hay	2 tons	80	8.5	56.

Potassium is needed by all crops. It is especially needed in hay crops, in the straw of grain crops, and in potato and root crops. It is potassium that gives stiffness to the straw of grain crops and enables it to stand up and mature a crop. Where there is a scarcity of potassium the straw of grain crops is likely to be weak. Potassium is removed from the farm rapidly by such crops as hay, potatoes and sugar beets. The amount of potassium in ordinary soils varies from 20,000 to 50,000 lbs. per acre.

Calcium is used in a very limited extent by plants, as plant food, except by legumes, such as alfalfa, clover, peas, etc. Its most important use in the soil is to overcome acidity or sourness. When a soil is sour, bacteria necessary for the liberation of plant food are likely to be lacking, especially those bacteria associated with the accumulation of nitrogen in the soils and its conversion into forms available for plants. A sour soil is often indicated by the growth of such weeds as sorrel and horsetail fern, or by its failure to produce good crops of legumes. It is also easily detected by several simple tests. Many soils have an abundance of lime, while others are seriously in need of it. The amount of calcium in normal soils varies from a few hundred pounds to a great many tons per acre.

Questions:
1. What do you understand by the term fertilizer? By commercial fertilizers?
2. Name the four elements that are likely to be lacking in cultivated soils, and tell at least one special purpose for which each is needed.

Arithmetic:
1. If a soil contains 1,000 lbs. of phosphorus per acre, how many 100-bu. crops of corn can it produce, without the addition of more phosphorus? If 1 bu. of corn removes 12-100 lbs. of phosphorus?
2. If nitrogen is worth 18c. per lb. as a fertilizer, what is the value of 2 tons of clover hay to plow under? If it contains 2 lbs. of nitrogen per 100 lbs.?
3. If a soil contains 50,000 lbs. of potassium per acre, how many 100-bu. crops of corn can it produce without exhausting the potassium, if 1 bu. of corn removes 2-10 lbs. of potassium?

FERTILIZERS AND THEIR USE

Costs.—We have learned that there are four elements needed by plants, nitrogen, phosphorus, potassium and calcium, the supply of which is likely to become depleted in ordinary soils unless care is taken to maintain them. In the last lesson we learned that a 20-bushel yield of wheat removed 22 lbs. of nitrogen, 4.1 lbs. of phosphorus and 5.8 lbs. of potassium. Nitrogen costs about 18 cents per pound, phosphorus about 6 cents per pound and potassium about 6 cents per pound, when purchased in the form of commercial fertilizers. To replace the elements removed by 20 bushels of wheat would cost at these prices about $4.55 or 22¾ cents per bushel of wheat. It it were necessary to purchase all these elements used, it would make a very heavy tax on wheat raising.

Nitrogen is present in the air in very large quantities. The farmer has a way at his command by which he can gather this nitrogen and add it to the soil at practically no cost. This is done by growing legumes, such as clover, alfalfa, etc. (See Chapter VI) A good rotation of crops (See Chapter XXIII) in which clover or some other legume crop is included one or more times in from three to seven years will provide for the maintenance of a sufficient supply of nitrogen. Barnyard manure also contains liberal amounts of nitrogen and its application to the soil every few years is very helpful in maintaining a supple of nitrogen. It is not necessary, therefore, under ordinary farm conditions, to purchase nitrogen. It is very important in connection with the maintenance of nitrogen that a liberal supply of vegetable matter be maintained. When it becomes necessary to purchase nitrogen for a fertilizer, it may be secured in the form of sodium nitrate, a product taken from mines, ammonium sulphate, a by-product in the manufacture of gas, dried blood and other by-products from large packing plants or slaughter houses.

Phosphorus.—It will be noted in the last lesson that phosphorus is present in the soil in comparatively small quantities. It is removed from the farm when either grain or live stock is sold. The addition of manure adds to the supply, especially if mill feed is purchased. Often the elements of fertility may be more economically purchased in the form of feed to be fed to live stock and the manure applied to the soil than to purchase them in the form of commercial fertilizers. Maintaining a supply of vegetable matter in the soil is important in aiding the liberation of phosphorus contained in the soil. Even after one has taken all these precautions to conserve phosphorus, there is almost certain to come a time when the supply of available phosphorus in the soil will be so small as to seriously limit the yield of crops. This is one of the elements that will no doubt have to be furnished to the soil in the form of commercial fertilizers. The more important sources of phosphorus fertilizers are ground bones, either steamed or raw from slaughter houses, and mineral phosphate mined

Figure 14.—Making use of straw to maintain fertility instead of burning it.

in many places in the United States. This is most commonly used in a form called acid phosphate. In this form the phosphorus is easily available for plants. Raw, finely ground phosphate rock is now used in quite large quantities. This is a cheaper form of phosphorus, and, where a good supply of vegetable matter is maintained, the decomposition of the vegetable matter in the soil renders the phosphorus in the raw rock soluble, so that the plants can use it. The finer the raw phosphorus rock is ground the more easily is the phosphorus made soluble.

Potassium is much more abundant in most soils than phosphorus or nitrogen. Most of the potassium used by plants is in the stems or straw. Most of the straw and hay produced on farms is used as feed or bedding for live stock and returned to the soil in the form of manure. For these reasons there are very few soils that are deficient in potassium. In fact, in general farming it is seldom necessary to use fertilizers containing potassium other than barnyard manure especially if a practical rotation of crops is followed. Occasionally it pays a farmer to apply potassium for hay crops, root crops and potatoes, as these crops use very large amounts of this element. When it is necessary to apply potassium fertilizers, the most common kinds sold commercially are some of the mineral potassium salts, such as kainit, muriate of potash, a product refined from the mineral potassium salts, and wood ashes.

Calcium is generally quite abundant, but occasionally there are soils that have very little or no calcium. Where there is insufficient calcium the soil is sour and crops do not succeed. This condition cannot be corrected by applying manure. Calcium in some form must be supplied. Calcium is commonly spoken of as lime. There are several different forms in which calcium may be applied. Common quicklime (the lime used for plastering) may be used, or finely ground limestone or slaked lime or marl will also accomplish the same result.

Complete Fertilizers.—There are many companies that prepare fertilizers ready for use on the farm. These fertilizers usually contain some calcium, some nitrogen, some phosphorus, and some potassium, also a considerable amount of other material called filler to make up the bulk and weight. Such fertilizers are called complete fertilizers. It is seldom economical to use such fertilizers, because it is seldom necessary to apply all of the elements to a soil and the elements not needed, if applied, are wasted. As stated above, the cheapest and most practical way of adding nitrogen to the soil is to grow clover and other legume crops. Most soils, in fact nearly all soils, have enough potassium, so that it is not necessary to add more. Phosphorus is the element most likely to be needed and the cheapest way to supply that is to buy a fertilizer containing only phosphorus.

Animal Manure.—Except in the case of soils that may be seriously impoverished of some particular element of plant food, it has been found that the total crop products from fields that have received animal manure, exceed those from land treated with commercial fertilizers or which were not manured. Animal manure is preferable because

(1) it is produced at a minimum of expense,
(2) it adds immediately available plant food to the soil,
(3) it provides humus and acid solvents that assist soil decay,
(4) it produces effects for years after application.

The Composition and Amounts of Manure Produced by Different Kinds of Farm Animals is Shown in the Following Table:

Kind of Animal and Kinds of Food Fed	Analysis				Amount per 1,000 lbs. Live Weight		
	Per Cent Water	Per Cent Nitrogen	Per Cent Phosphorus	Per Cent Potassium	Lbs. a Day	Lbs. a Year	Total Tons Manure per Year
Cattle. Fed hay, silage, beets, wheat, bran, cornmeal, cottonseed meal	75.25	.43	.127	.44	74.1	27,046	15.0
Horses. Fed hay, oats, cornmeal, wheat bran	48.69	.49	.114	.48	48.8	17,812	10.5
Sheep. Fed hay, corn, oats or hay, wheat bran, cottonseed meal, linseed meal	59.52	.77	4.10	.59	34.1	12,446	8.7
Swine. Fed skim milk, cornmeal, meat scraps; or cornmeal, wheat bran, linseed meal	74.13	.84	.17	.32	83.6	30,514	17.7

The data in the above table are taken from the Cornell Experiment Station.

Questions:

1. Describe the most practical means of maintaining a supply of nitrogen in a soil for ordinary farming?
2. Which of the four elements named in this lesson is most likely to become exhausted? How may it be replenished?
3. Is much potash removed from soil with ordinary crops? Why?
4. What is a complete fertilizer? Is it usually economical?
5. What are the advantages of animal manures?

Arithmetic:

1. If 50 bus. of corn remove 31 lbs. of nitrogen, 6.1 lbs. of phosphorus, and 9.6 lbs. of potassium, how much would it cost to replace these elements at 18c. per lb. for nitrogen, and 6c. per lb. for potassium and phosphorus?
2. If an acre of normal soil contains 4,000 lbs. of nitrogen, 2,000 lbs. of phosphorus and 35,000 lbs. of potassium, what is the total value of these elements per acre at prices used in Example 1?
3. If acid phosphate contains 6% phosphorus and a 50-bu. crop removes 6.1 lbs. of phosphorus, how many pounds of acid phosphate would one have to apply to supply the needs of the crop?

CHAPTER IV

GRAIN CROPS

PLANT STRUCTURE

Parts of Plant.—All the more important plants, in which the farmer is chiefly interested, have four distinct parts, roots, stems, leaves, and flowers. Plants, like animals, vary greatly. This fact is true of plants in any one variety, such as wheat plants or pansy plants. This habit of variation has made possible the development of the great variety of plants that are grown to supply the many different needs of man. In some plants one part has been developed for use, in another plant other parts. For example, in the turnips the root is eaten; in asparagus the stems; in lettuce the leaves; and in wheat the seeds.

Roots.—There are two kinds of roots, fibrous roots and tap roots. Roots grow in the soil and take up moisture and plant food. The moisture in the soil and the plant food with it pass through the thin walls of the roots and very fine root hairs, or branch roots. Roots also hold the plant in place and tend to keep it erect.

Stems.—These may grow erect like corn or wheat or may trail along the ground like squash or cucumber vines, or they may climb up some other object, like peas or beans. The function or work of the stem is to bear the leaves and blossoms and to provide a means for the plant food to go to and from the leaves. Plant food and moisture circulate in the plant somewhat similar to the circulation of blood in the body of an animal. In some cases, like potatoes, for example, some of the stems grow under ground; in fact, the potato (tuber) that we eat is simply an enlarged stem. This fact is indicated by the eyes on the potato, which are the buds from which branches may grow.

Leaves have been called the stomach of the plant, because it is in the leaves that the plant food from the soil and from the air are brought together and changed into the compounds which make up the plant. From 300 to

500 lbs. of moisture must pass through a plant to produce one pound of dry matter in the plant. This water must all be given off into the air by the leaves, hence the need of so many leaves. A tree, or a plant, usually has a very large leaf area exposed. It is interesting to estimate this leaf area on various plants, by counting the leaves on a part of the plant, measuring the size of an average leaf and then computing the total, remembering that there are two sides to each leaf. Each leaf has many little openings or pores on its under surface through which air is taken into the leaf and moisture and oxygen given off.

Flowers are borne on the stems of plants. Their function is to start and develop the seed which enables the plant to reproduce itself. In fact, the production of seed seems to be the function of the whole plant, but it is the blossom that starts the seed and as soon as the seed is started the blossom falls.

Questions:
1. Name the four principal parts of a plant.
2. Name at least two plants in which the edible part comes from the roots. From the stem? From the leaves? From the seed?
3. Tell the principal uses to plants of roots and leaves.

Arithmetic:
1. If 400 lbs. of water are given off (transpired) by clover plants to produce 1 lb. of dry matter, how many pounds will be transpired from an acre of clover yielding 3 tons of hay containing 85% of dry matter?
2. Count the leaves on an average sized hill of potatoes. Measure an average sized leaf and find the total leaf surface exposed per hill. Per acre. If a potato plant is not available, use a geranium plant instead.

GOOD SEED

Importance.—We must have good kernels of grain from which to raise a crop as good horses, sheep or cows from which to raise colts, lambs or calves. One of the laws of Nature which we must consider in raising plants and animals is that "Like produces like." If we want to raise large horses we must have large horses from which to raise them. If we want to raise dairy cows we must keep dairy cows or cows which have the ability to produce large amounts of milk. Likewise, if we wish to produce good plants, we must sow good seed. There are small, large, shrunken and plump grains. It is important to know which we should plant.

Test of Good Seed.—Good seed of any kind of grain must have at least three qualities:

(1) It must be pure, that is, free from weed and other grain seed.

(2) It must be well matured, plum and heavy.

(3) It must germinate well so as to produce strong plants.

You will notice by examining a small sample of grain (place a small sample on a piece of white paper) that there is a great difference in the size, character and shape of the kernels. (Separate the sample into good and poor lots.) Would you care to plant the poor seed? You might be interested to plant ten of the very best, large, heavy seeds and ten of the poorest, small, light seeds in a box of pure sand. Moisten the sand and keep the box in a warm room. See which seeds produce the larger, stronger plants.

Parts of a Seed.—A seed is made up of three parts:

Corn Kernel. (1) A small plantlet or germ, the embryo, inside of each kernel which will, when the seed is placed under favorable conditions as to heat, air and moisture, grow and produce a plant;

(2) The food material stored about the embryo, to feed it until it has developed a root system so as to be able to get food from the soil; and

(3) The seed coat on the outside for protection.

It is evident that a large, plump kernel or seed will have a stronger, larger germ than will a small or shrunken seed, and will also have more food for the little plantlet, so the plantlet will get a better start before it must obtain its food from the soil.

Select Seed from Best Plants.—Another reason for selecting the large, plump seeds is because it is reasonable to expect that they grew on good, strong, healthy plants. There are a great many unfavorable conditions with which plants have to contend, such as diseases like smut, rust and blight; unfavorable weather conditions, as cold or wet or drought or heat; also poor soil conditions. It is evident that some plants are better able to withstand such conditions than others. Those that do withstand such unfavorable conditions and are best adapted to the soil and climate will be likely to do better and produce better and more perfect seed than will other plants. The heaviest and plumpest seeds are selected when seeds from the best and most vigorous plants are secured. Hence, when a farmer selects the heavy, plump seeds raised on his own farm, he not only gets good, strong seed, but seed adapted to his soil and climate.

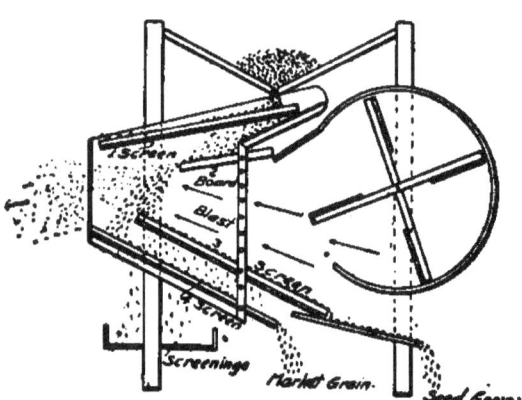

Figure 15.—Diagram of a Fanning Mill, showing a good method of grading seed grain. The blast blows the lighter kernels over the end of screen No. 3, in with the market grain. The heavy kernels fall on this screen. The smaller of these go through into market grain, leaving only the heavy large kernels to go into the seed grain.

Heavy seed grown in some other part of the country is usually heavy because it grew under favorable conditions rather than because it came from especially strong plants. The heaviest home-grown seed is often better to sow than still heavier seed from other localities.

Questions:
1. What law of Nature must be considered in raising plants and animals?
2. What are the three qualities that all good seed must possess?
3. What two kinds of seed can you usually find, if you examine a good, pure sample of grain?
4. What are the three parts of a seed, and the purpose of each?
5. Why is heavy, plump seed better than small or shrunken?

Arithmetic:
1. Land at the Minnesota Experiment Station, seeded with heavy, plump seed oats, yielded 9.5 bus. per acre more than similar land seeded to light weight oats. What was the additional income per acre from heavy weight seed if oats are worth 38c. per bushel?
2. If 2 bus. of heavy seed oats (sown on one acre) give an increased yield of 9.5 bus. what would be the increased yield from one bushel of heavy weight seed oats?
3. If one bushel of heavy weight seed oats gives a yield of 4.5 bus. more than is secured from a bushel of light seed, what is the value of one bushel of heavy weight seed when oats are worth 38c. per bushel?

Note: If the light weight seed is worth 38c. per bushel, the heavy graded seed will be worth 38c. plus 4.5 times 38c.

SELECTION OF GOOD SEED

Pure Seed.—It is well worth while for farmers to raise only pure varieties of grain, or grain that contains no other kind or variety of seed. Seed of Blue Stem wheat should be free from oats, rye, and barley as well as from other kinds of wheat. Pure seed grain may be secured by purchasing a small amount of pure seed and using care in growing it so that it will not become mixed; or, if one prefers to start with the seed on the farm, one may go through a small patch of grain when it is headed out in the field and pick out and destroy the other kinds of grain, thus getting a small patch pure, from which a start in pure seed may be made.

As a rule the very best seed that one can get for the main crop is from grain that has been grown on the farm for several years and that has given good yields. Such grain, when graded and cleaned, so that only the very best is saved for seed, usually gives excellent seed.

The two general principles by which grain can be graded, separated or cleaned of weed seeds, by the use of the fanning mill, are by size and shape of kernels or by weight.

To Remove Weed Seeds.—Most weed seeds may be removed from grain by running the grain through a fanning mill. The large weed seeds are separated from the grain by dropping the grain through a sieve that is too fine to let the weed seeds through. The small weed seeds are taken out by running the grain over a sieve that is too fine to let the grain through but coarse enough to let the small weed seeds through. The weed seeds that are lighter than the grain may be blown out. Sometimes the light grains, like oats, may be separated from heavy weed seeds by blowing the grain out of the weed seeds.

There are some weed seeds, like cockle and wild vetch, which are about the same size and weight as wheat, that are very hard to separate from that grain. While such weed seeds as wild oats are very hard to separate from oats and barley, as the seeds are quite similar in character. When such weed seeds are present in grain and can not be removed with a fanning mill, a small amount of seed free from such weed seeds may be secured by hand picking or by pulling the weed plants from a small plat of growing grain, thereby getting a start in clean seed.

Grading Seed Grain.—Many persons are satisfied when they get pure and clean seed grain; but, if one wishes to get the best results and maintain or improve grain from year to year, it is necessary to grade out and use for seed only the very best individual seeds in the whole amount grown on the farm. This may be cheaply done by grading the grain as shown on page 42. In this way the heavy plump, kernels are separated from the smaller, lighter ones. The former kernels should be used for seed and the latter sold or used for feed.

Germination.—It is as important that seed grain germinates (starts to grow) well as that seed corn germinates well. It is a very simple matter to test seed grain for germination, and this should always be done before it is planted. A good germinator is made as follows: Partly fill a plate with sawdust or sand, cover with a cloth, and on this scatter one hundred seeds. Cover with another cloth, moisten and cover all with an inverted plate. See Figure 16.

Figure 16.—A simple germinator for testing seed grain. The lower plate is partly filled with sand, the grain placed between the cloths on top of the sand and all covered by an inverted plate. If kept moist and in a warm room, good seed will germinate in from 5 to 7 days.

Another simple germinator for grain and grass seed is made out of blotting paper. Place the seeds to germinate between two squares of the blotting paper. Good seed in such a germinator, kept moist and in a warm room, as a living room or school house, for from five to seven days, will sprout. The number out of the one hundred that start to grow vigorously in that length of time will represent the per cent of the seed that will be likely to grow in the field. It will pay to try this with several different kinds of grains.

Questions:
1. From what source is one most likely to get good seed grain for most of a planting? Why?
2. In what ways may weed seeds be separated from seed grain?

3. By what method can one cheaply separate the large, plump kernels of grain from the small and light kernels?

Arithmetic:

1. Two men can clean and grade with a fanning mill 20 bus. of grain in an hour. If the best 30% is saved for seed, how many bushels of seed will be secured? How much will it cost per bushel to save seed in this manner, if each man's time is worth 15c. per hour?

2. If it costs 5c. per bushel to grade out the best seed and it requires two bushels of seed to seed an acre, how much must the yield be increased per acre to pay for the extra cost of grading the seed, if oats are worth 38c. per bushel?

3. If one seeds 50 acres to grain that germinates but 80%, what proportion of the land is seeded to grain that will not grow?

WHEAT

Importance.—Wheat is the chief source of bread commonly eaten in the United States; in fact, in the greater part of the world. White bread is found on nearly every table at every meal. It is even more common than potatoes. The United States produces about 700,000,000 bushels of wheat annually. It is one of the large wheat-producing countries of the world. Russia produces about the same amount. Each person in America uses about 4.7 bushels of wheat per year. The United States exports, on the average, over 100,000,000 bushels each year, either in the form of wheat or flour. The central western states, from Ohio and Indiana west to Kansas and the Dakotas, are the principal wheat-producing states.

The wheat yield of the United States for 1914 was 878,680,000 bushels valued at $610,122,000. This was the largest crop ever raised in this country. The average per acre was raised from 16.7 to 19 bushels.

Place.—Wheat has been grown in the central western states quite generally for the first years of their development. In the older states it is usually replaced to a considerable extent with corn and grass, and, while there is still a large acreage of wheat grown in these states, it is by no means the chief crop. It has a very important place on a general farm, and, when grown in rotation with other crops, it is a profitable crop. Wheat ordinarily does not yield as large a money return per acre as corn or potatoes, but it may yield as much profit, because a comparatively small amount of labor and expense are required to grow it. Wheat is one of the best grain crops to use as a nurse crop for clover, timothy and other grasses.

Kinds of Wheat.—There are a great many different kinds of wheat; but, so far as most farmers are concerned, there are but two general types; namely, spring wheat and winter wheat. Spring wheat is seeded in the spring and harvested in late summer. Winter wheat is seeded in the fall, lives over winter, and is harvested in midsummer. Minnesota and the Dakotas are the principal spring wheat-producing states. Most of the other important wheat states produce winter wheat.

Soil for Wheat.—Wheat will do well on any ordinarily productive soil in the United States. It does best fol-

Figure 17.—A fine field of wheat.

lowing some cultivated crop like corn or potatoes, or on land that has been summer fallowed. Land is usually plowed to a medium depth of from four to six inches. Fall plowing is preferred for spring wheat. The soil is then harrowed or disked, or both, until the top two or three inches are mellow and fine.

Sowing.—Wheat is sown at the rate of one bushel to one and a half bushels per acre. It may be sown in drills, usually six inches apart, or broadcast. Drilling is preferred, as it places the seed in the soil at an even depth, and it is all covered.

Harvesting.—Harvesting wheat used to be a very hard task; now it is done by machinery and very rapidly. In the central and eastern part of the United States it is harvested chiefly with binders. A man with three or four horses and a binder can cut and bind from ten to eighteen acres per day. In western part of the United States, where large acreages of wheat are grown, much of the harvesting is done with large combination machines that cut the wheat, thresh it and sack it in one operation. These machines are hauled by large traction engines, or by thirty or more horses. Such machines can be used only where there is no danger of rain, so the wheat can stand until thoroughly ripe.

Shocking and stacking of wheat are done by hand. A good shock of wheat must provide that the heads of all bundles be kept off the ground and protected from rain. This is done by setting eight or more bundles firmly on the ground, butt end down, and leaning them against each other so that they will stand erect. The shock is then capped or covered with one or two bundles, so placed as to shed water and protect the heads of grain from sun and dew. Stacking is quite an art, and men pride themselves in building uniform, straight stacks that will not blow over, and that will protect all the heads of grain from sun and rain.

Figure 18.—A good shock of wheat well capped.

Threshing usually results in a very busy time on the average farm. A man in each community usually owns a threshing outfit, consisting of an engine and a separator, which are moved from farm to farm. A modern outfit will thresh from 1,500 to 4,000 bushels of grain in a day.

Questions:
1. Tell what you can about the United States as a wheat-producing country.
2. Tell what you can about the growing of wheat.
3. Which would you use, a drill or a broadcast seeder in sowing wheat? Why?

Arithmetic:

1. What is the value of an acre of wheat yielding 13.5 bus. at 85c. per bushel?
2. If it costs $12.60 per acre to produce wheat, how much does it cost per bushel, if wheat yields 13.5 bus. per acre? If it yields 20 bus. per acre?
3. If a man is sowing wheat with a drill 8 feet wide, how many miles will he have to travel to sow 80 acres?

OATS

Importance.—The importance of oats in the United States is shown by the fact that about one billion bushels are grown each year. The important oat-producing states are in the central west, with Iowa, Illinois, Wisconsin, Minnesota and Nebraska leading. Oats are commonly grown throughout the world, and in the same countries that produce wheat. The United States, Russia, Germany, France and Canada are the principal oat-producing countries.

In 1914 there were raised in the United States, 1,141,060,000 bushels, valued at $499,431,000. The acreage was 38,442,000.

Uses.—The chief use of oats is for grain feed for horses, and oats are by far the most popular feed among horsemen. Oats also produce the best and most practical breakfast food known. They are used, and are very satisfactory, for feed for all other classes of stock, but the demand for them in the cities for horse feed usually makes the selling price for them so high that they are not fed to any great extent to any stock on the farm but horses.

Figure 19. A good sheaf of oats well-capped.

Culture.—Oats are grown in a manner very similar to that in which wheat is grown. Most of the oats grown in the United States are of the spring varieties, because the winter varieties are not hardy except in the southern and extreme western states. Oats will do well on a great variety of soils, just the same as wheat. Fall-plowed land is preferred, but spring plowing will do. A large proportion of the oat crop of the United States is sown on land that produced corn the year before. Such land is usually not plowed, but is disked thoroughly instead. Oats are sown early in the spring at the rate of from two to three bushels of seed per acre. The crop is usually ready to harvest in from eighty to one hundred and twenty days from the time it is sown. Oats are harvested, stacked and threshed in the same manner as wheat is.

BARLEY

Barley is not grown in nearly so large quantities as oats or wheat. Russia is the chief barley-producing country. The United States, Austria-Hungary, Great Britain, Germany, Canada and Spain are other countries that produce barley in large quantities.

Barley is used for malting, that is to make beer, and for feed. It yields more pounds per acre than oats, but is not so popular as a feed as oats, and the crop is a little more unpleasant to handle than the other grain crops on account of the beards or awns. Barley, however, is a good feed for all classes of stock, and is used quite generally as a substitute for corn where corn is not easily grown. It is an early maturing, heavy growing crop, and on that account is one of the best spring-sown grain crops for cleaning the land of weeds. There are several different types of barley: the six-row and two-row bearded, the hull-less and beardless. By far the most important type is the six-row bearded.

Culture.—Barley does best on rather rich soil. On light soils the straw is likely to be so short that it is difficult to cut with a binder. Barley is the most tender to frost of any of the grain crops. Frost in the spring will quite seriously injure barley, while wheat, rye and oats are unharmed. On this account the crop is usually sown from two to four weeks later in the spring than other grain crops. From seven to eight pecks of seed are sown per acre.

Harvesting.—Barley is cut and shocked in the same manner as wheat. Special care must be taken with barley,

Figure 20.—Some good grain stacks.

if it is to be sold for malting, to protect it from the weather, as bleaching of the kernels materially reduces the value for that purpose. It must, therefore, be cut as soon as it it ripe, carefully shocked in capped shocks, and stacked or threshed as soon as it is dry enough.

RYE

Rye is the least important cereal crop in the United States, but is more important as a world crop than barley. Russia produces the largest acreage of rye of any country. Germany, Austria-Hungary, Norway, and Sweden, and France are other important rye-producing countries. In these countries it is used to a much greater extent for bread than in the United States. It is used for bread, for the manufacture of alcohol, and as feed for stock. The green crop is often used for pasture, and it will furnish pasture earlier in the spring than most other crops. Rye is an excellent crop to grow in cleaning land of weeds, because it matures early.

Culture.—There are two types of rye, spring and winter. Winter rye is the more common in the United States. It is very hardy, and will grow on almost any kind of soil. It is commonly sown on the lighter soils, because it will do better on such soils than other grain crops. It is usually sown

GRAIN CROPS

in the fall from August to October, at the rate of five to six pecks per acre. It is harvested in about the same manner as other grain crops are.

Questions:

1. Tell what you can about the importance, uses, varieties and culture of oats, of barley, of rye.

Arithmetic:

1. If rye yields 16 bus. per acre, weighing 56 lbs. per bushel, how many pounds are produced per acre?

2. If oats yields 30 bus. per acre, weighing 32 lbs. per bushel, how many pounds are produced per acre?

3. If barley yields 25 bus. per acre, weighing 48 lbs. per bushel, how many pounds are produced per acre?

CHAPTER V

CULTIVATED CROPS

CORN

GENERAL FEATURES OF THE CORN CROP

Corn is an odd but true grass like timothy or wheat. It is native to America and was imported into Europe by Columbus who found it cultivated by the Indians when he discovered this continent. It is adapted to temperate zones but may be acclimated to the northern regions of the United States. It is grown most in the United States, Austria-Hungary, Argentine, Russia, Egypt and Australasia.

Importance.—Corn is more widely cultivated, yields a larger crop than any of the other cereals, and its total value is greater than that of any other crop grown in this country. In 1914 the United States produced its largest crop—2,672,804,000 bushels, valued at $1,702,599,000. The average acre yield was almost three bushels more than in 1913. The average number of bushels produced in the world for the years 1905-1909 was 3,585,418,600, of which 76% was produced in the United States. The average acre value of corn in the United States is $12.53. The value of the grain alone in this country is greater than that of any other two farm crops produced and greater than that of the wheat, oat, barley, flax, rye and tobacco crops combined. The grain is used in different forms as food for both man and beast, while its many manufactured and by-products have extensive uses. The fodder is also made to serve in various ways as food for stock.

Yield of Corn.—Average yields of corn in even the best corn-growing states of the Union are very low, much lower in fact than yields secured by the best farmers. It is well worth the time of any one interested in farming to know the methods practiced by the best farmers, that the maximum yields at the least possible expense of labor and fertility may be secured.

As farmers change from grain raising to a more diversified type of agriculture, more live stock will be kept and more corn raised. The average 160-acre farm will then raise from 30 to 50 acres of corn each year.

A Young Man's Opportunity.—If a young man begins farming on one of these farms when he is twenty years old, and continues until he is fifty, he will raise during his active life approximately 1,200 acres of corn. It will make quite a difference to him and his family whether he follows indifferent methods of farming and gets an average yield of 30 bushels per acre, or whether he follows good methods and gets 50 bushels per acre. It will pay, and pay well, any boy who expects to raise corn to thoroughly master the subject, so that he will get the extra 20 bushels per acre.

Requirements.—Nearly every large business is made up of many details, and corn growing is no exception. The four general requirements for a good crop of corn are, good seed, good soil, good tillage and good climatic conditions.

Good seed is easily secured by selecting good ears of corn from good plants, and by carefully curing, storing, testing and grading it.

Good soil may be had in almost any part of the United States by properly caring for the land we have. By practicing rotation of crops, by keeping live stock and feeding on the farm most of the field crops raised instead of selling them and thus losing fertility, by draining land that is too wet, and by keeping in check noxious weeds, land may be maintained at a high state of productivity.

Good tillage means doing the things which make the soil the best possible place for the crop to grow. This requires a knowledge of the soil, of the movement of water in the soil, of the habits of plants, and of the methods by which plant food is liberated.

Climatic Conditions.—Good seed, good soil, and good tillage are within the control of the farmer. Climatic conditions are not, though he may do many things to guard against unfavorable weather. He can drain his land to avoid an excess of moisture and to make his soil warmer. He can regulate his tillage operations to conserve moisture in case of drouth and to aid in warming the soil, if it is too cold. By manuring land and by growing clover occasionally, he can make a soil warmer, more retentive of moisture, and increase its producing power, so that crops will grow more rapidly and thus ripen in a shorter time. Climatic conditions are usually favorable, so that, with good methods of farming, good crops can be grown practically every year.

Some farmers in the corn belt have raised more than one hundred bushels of corn per acre. Let us set our standard at that and be satisfied with nothing less.

Questions:
1. Why do you think it worth while for a boy to study about corn growing?
2. Name at least four conditions necessary to secure a good crop of corn.
3. What may one do to reduce the bad effects of unfavorable weather?

Arithmetic:
1. A and B each grow 40 acres of corn per year for 30 years. How many acres does each grow in the 30 years?
2. If B uses the best known methods of corn growing and secures an average yield of 20 bus. per acre more than A, how many bushels more corn will he raise in 30 years than A raises? How much will his extra corn be worth at an average price of 54c. per bushel?

SIZES OF KERNELS OF CORN

Variation of Corn Kernels.—There is a great variation in the size of kernels of corn; and while this has little to do with the yield—that is, some varieties with comparatively small kernels may yield more than other varieties with large kernels—it is, nevertheless, important to select ears of corn on which the kernels are about uniform in size.

To Compare Kernels.—Shell kernels from the tip, the butt and some from the middle of an ear of corn, keeping the three kinds separate. Lay three of the even sized kernels from the middle of the ear together on a sheet of paper and draw a circle about them. Make the circle just large enough so the three kernels will lie flat within. (Circle about the size of a nickel.) See how many of the small tip kernels can be laid in this circle and how many butt kernels. This circle is about the size of a hole in the plate of the corn planter. Notice how much thicker some of the butt kernels are than kernels from the middle of the ear.

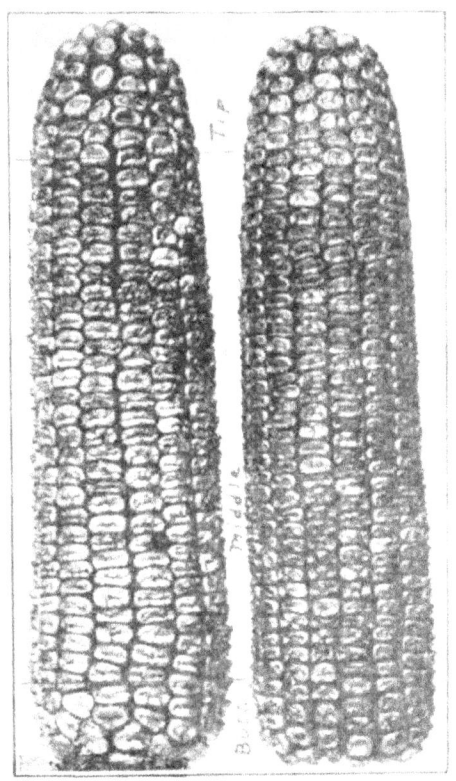 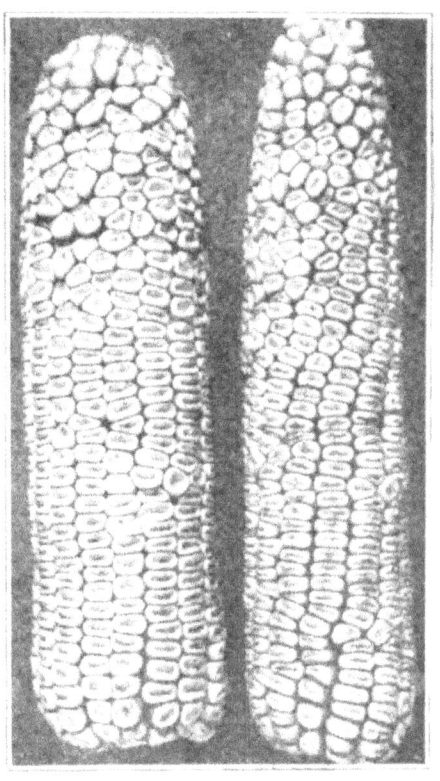

Figure 21.—Good ears of corn, with straight rows and even kernels.

Figure 22.—Poor ears of corn. Note the crooked rows and irregular kernels.

Examine a corn planter, if possible, to see how it drops the corn.

It is very important to the farmer that all his seed corn be uniform in size, because corn is now planted by machines and unless the kernels are about the same size and shape the machine cannot drop the same number in each hill. If uneven sized kernels were used for planting, the number in a hill would vary as the number of kernels you were able to place in the circles you drew varied.

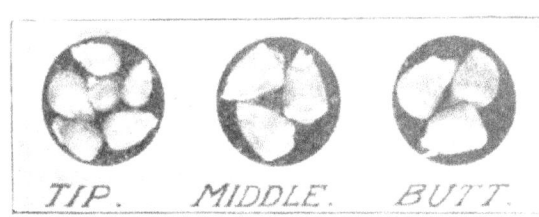

Figure 23.—Relative size and shape of tip, middle and butt kernels of corn. The circle represents the hole in the plate in a corn planter, with the number of kernels of different sizes that a planter would drop. Photo by H. D. Ayer.

Most farmers like to plant three kernels in a hill, because

they have found that three stalks to a hill give the best yields.

Even Seed.—Farmers can get even corn to plant by selecting even, straight rowed ears of corn, and by shelling off the tip and butt kernels, using this part for feed and saving only the more even kernels from the middle of the ears to plant. The whole crop on an acre of corn depends on a few ears of seed corn.

Questions:
1. Do the size and shape of kernels of seed corn make any difference to the farmer?
2. How can farmers get even seed corn?
3. If the tip kernels were put into a planter, would it drop too many or too few?

Arithmetic:
1. After the tip and butt kernels of corn have been shelled off from an ear, count the number of rows of kernels; then count the number of kernels in one row. How many kernels on the ear of corn?
2. Find how many hills of corn on an acre when corn is planted in hills 3 ft. 8 in. apart each way.
Note.—There are 160 sq rds. in an acre, and each hill of corn takes up 3 ft. 8 in. x 3 ft. 8 in. or 13 4-9 sq. ft. of space.
3. If three kernels are planted in each hill, how many ears of corn like the one you counted are required to plant an acre?

PARTS OF A KERNEL OF CORN

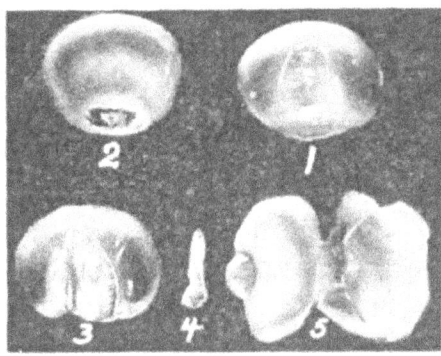

Figure 24.—Parts of a kernel of corn. 1. Side on which the embryo is found. 2. Side opposite the embryo. 3. Kernel with embryo and seed coat removed. 4. The embryo. 5. Seed coat removed from the kernel.

Examining Kernels.—It is not always easy to believe that there is a quite complete, though small, corn plant in each kernel of corn. If you will dissect a few kernels of corn, they will furnish a very good object-lesson. A kernel of corn consists of three parts—an outside shell or seed coat, a little speck of life, or the embryo, and about the embryo a white, starchy substance or food portion.

The seed coat may be easily removed from a kernel of corn that has been soaked for a few minutes in hot water. It is hard and tough. Its purpose is to protect the parts within. It protects them from heat, cold, and moisture.

The Embryo.—The speck of life, or embryo, may be taken from a soaked kernel of corn by the use of a sharp knife or a needle. It is really a very tiny live corn plant, and is found bedded near the tip of the kernel, in the white starchy part. This embryo has a root and a stem. The stem is not green, however, because it has been shut up in the dark. If corn is properly ripened and kept dry, the little plant within each kernel will stay alive a long time, and be ready to grow when the kernel containing it is put into the ground and supplied with heat and moisture. If corn be allowed to freeze, perhaps thaw out and freeze again, the little embryo within a kernel is not likely to live, and the kernel would not grow if planted. It is for this reason that farmers are careful to select for seed well ripened ears of corn, and to store them safely in a dry place.

Food Material.—After the seed coat has been removed and the embryo taken out, a large part of the kernel is still left. It is the food part. It makes food for us when the corn is ground into meal; or, when the seed is planted and the embryo begins to grow, it is this part which furnishes the embryo with food until it has developed roots and can get its food from the soil. The large kernels have more of this food material than the small ones. A stronger plant will grow from a large kernel than from a small one, on account of the greater amount of food material the larger kernel contains for the early growth of the little plant. This is one reason why plump kernels are better for seed.

Questions:
1. A kernel of corn consists of what parts?
2. Tell all you can about each part.
3. Where in the kernel did you find the embryo?
4. What would injure or kill the embryo?
5. How should corn which you mean to plant be kept?
6. From which kernels come the strongest plants? Why?

Arithmetic:
1. If corn is planted May 15th and is struck by a frost Sept. 1st, how many days will it have in which to mature?
2. A bushel of seed corn will plant seven acres in check rows and is worth $2.50 per bushel. What is the cost of seed corn per acre?
3. A pays $5.00 per bushel for seed corn, B pays $2.00 per bushel.

Each one plants seven acres with his bushel of seed. How much more corn must A get per acre than B to pay the extra amount for his seed, if corn the following fall is worth 54c. per bushel?

TESTING SEED CORN FOR GERMINATION

Germination.—A seed is said to germinate when it sprouts or begins to grow. Most farmers are careful to use kernels from the middle of the ears of corn, because the kernels are more even in size and shape and the corn planter can, therefore, drop the required number, usually three, to every hill. Suppose one ear of corn which has five hundred kernels has been frozen or otherwise injured so that the embryo in each kernel is dead. If the corn planter drops one of these bad kernels with two good ones in every hill until the five hundred bad kernels are all planted, there will be five hundred hills each with one stalk missing. This fault would reduce a farmer's yield; and the more of such ears he planted, the greater would be the reduction of his yield. If, on the other hand, all the seeds dropped in every hill were seeds that would grow, the farmer could be sure of a good stand of corn. This point is important, because it costs as much to prepare the land, plant and cultivate the crop for a poor stand as for a good one.

Figure 25.—A simple germinator consisting of a plate partly filled with sand, a cloth marked in squares for the corn from each ear to be tested, and the cloth and plate with which to cover the corn.

Will It Grow?— One cannot always tell by looking at an ear of corn whether or not the kernels will grow. A farmer, to make sure he is planting only good seed, must test his corn. He may test one hundred kernels taken at random from a number of ears or a sack of corn; but, if he finds that only 80 per cent of his corn will grow, he must use this poor seed or buy seed. A much safer and a very easy and simple way is to test each ear before it is shelled. One wishes to know if all or most of the kernels on an ear of corn will sprout or grow. If he takes ten kernels from one ear, and finds that all of the ten kernels sprout, he can safely assume that the rest of the kernels on that ear will grow. That is a good ear for him to plant.

If he takes ten kernels from another ear, and finds that none or less than half of them sprout, he rightly assumes that the rest of the kernels on that ear would not be likely to grow. That is not a safe ear to plant.

To Test Corn.—By testing each ear a person may throw out the poor ones and save the good ones, which enables him to use his own seed and to be sure of planting only good seed. A simple germinator may be made as follows: On a piece of white outing flannel draw with a soft lead pencil a six-inch square, and mark it off into nine two-inch squares, numbering the small squares from one to nine. Place the cloth thus marked over a plate of sand or dirt. The next step is to number nine ears of corn. This is easily done by fastening a small tag to the butt of each ear of corn with a pin, as shown in Figure 26. Take ten

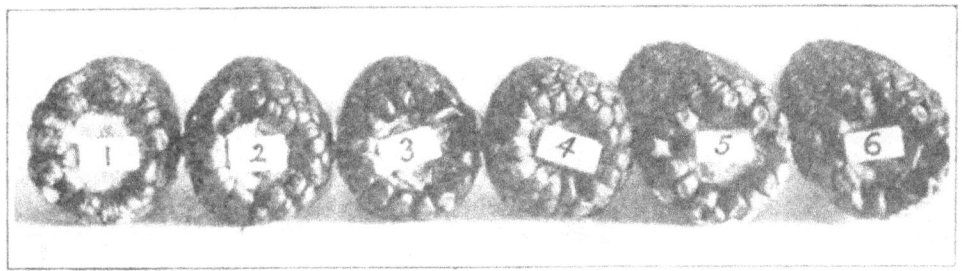

Figure 26.—Ears of seed corn numbered for testing.

kernels from ear No. 1, selecting them from different parts of the ear, and place them on square No. 1. Continue until each square is supplied with ten kernels from the ear of the corresponding number. Moisten the material in the plate thoroughly, cover with another cloth and another plate, to prevent too rapid evaporation, and set in a warm place. Put up carefully the nine ears of corn for future comparison.

The germinator should be examined from time to time to note the germination of the kernels. If kept warm and moist the corn should all germinate in five to eight days.

Kernels slow in germinating should be counted as worthless, as they would probably not grow in the field, if unfavorable conditions prevailed.

For testing a large amount of seed corn, as for ten to forty acres, a box two to four feet square may be used in place of the plates.

The Rag Doll Tester.—One of the most popular methods of testing seed corn is by means of the rag doll tester. These testers may be purchased ready-made, or may be made at home or at school. To make one use a good quality of muslin. Tear off a strip about nine inches wide and five feet long. With a soft lead pencil draw a mark lengthwise of the cloth and exactly in the center. Then every three inches draw lines crosswise of the cloth. Leave about fifteen inches on each end without crossmarks. Then number the squares, wet the cloth and the tester is ready for the corn. Number the ears of corn to be tested and place six kernels from ear number one in square number one. See Figure 27. Continue placing six kernels from each ear in a square of the same number until the tester is full. Place all kernels with the tips one way and pointing toward one side of the cloth. Then roll the cloth beginning at one end, using care not to displace the kernels. Tie a string about the roll and place it in a pail of warm water for a few hours. Then remove it from the water and place the roll on end in a pail or other dish eight or ten inches deep, and cover with a cloth and keep in a warm room. Place the rag doll in the pail so the tips of the kernels will point down. Several of the rag doll testers may

Figure 27.—A "rag doll" seed corn tester.

be placed in one pail. After from five to seven days the test may be read.

Questions:
1. For what reason should a farmer test his seed corn?
2. What are the advantages of testing each ear over testing 100 kernels out of a sack full of shelled corn?
3. How would you proceed to test 200 ears of corn?

Arithmetic:
1. If seven of the ten kernels taken from an ear of corn grow, what per cent does the ear germinate? If nine kernels grow, what per cent germinates?
2. If twenty ears of corn will plant one acre, what per cent of the corn in a field will be missing if the corn from one of the twenty ears will not grow? If the corn from three ears will not grow?
3. If a man test 400 ears of corn, and 90% of the ears are good enough to plant, how many acres of corn will the good seed plant? (Assume that twenty ears will plant an acre.)
4. A man can test 400 ears of corn in 6 hours. His time is worth 14c. per hour. If the 400 ears will plant 18 acres, how much does it cost him per acre to test his corn? If corn is worth 54c. per bushel in the fall, how much more corn per acre must a farmer get to pay him for thus testing his seed?

CORN CULTURE

The Corn Field.—In the spring of the year, when most farmers are preparing their fields for corn, will be a good time to study the planting phase of the corn subject. In the first place let us see on what kind of soil our neighbors and fathers and brothers are to plant corn. Land that produced clover or was pastured last year is best, as the clover and grass roots have filled the soil with vegetable matter, a very necessary condition for good crops. It would be better if the land were plowed last fall, as fall plowing gives the soil a chance to settle, so that it will not dry out readily.

Fall plowed land should be thoroughly disked in spring, before planting to corn, to insure a fine, mellow seed bed, to destroy weeds and to form a surface mulch to check the evaporation of water.

If there is no clover or pasture sod for corn, other well drained land, fall plowed, well manured and the manure thoroughly disked into the surface of the soil before planting, is the next best place for corn. The effort in any case should be to have a rich, firm soil, with about two inches

62 ELEMENTS OF FARM PRACTICE

of loose soil on top to check the evaporation of moisture.

Grade Seed Corn.—Practically all corn is now planted with a machine, and unless the kernels are of uniform size no machine can drop the same number of kernels in each hill, and it is important to do so.

As corn is shelled from the ear there is always more or less irregularity in the kernels. This is especially true if the tip and butt kernels are shelled with the rest. Even if they are not used, there are some irregular kernels in the

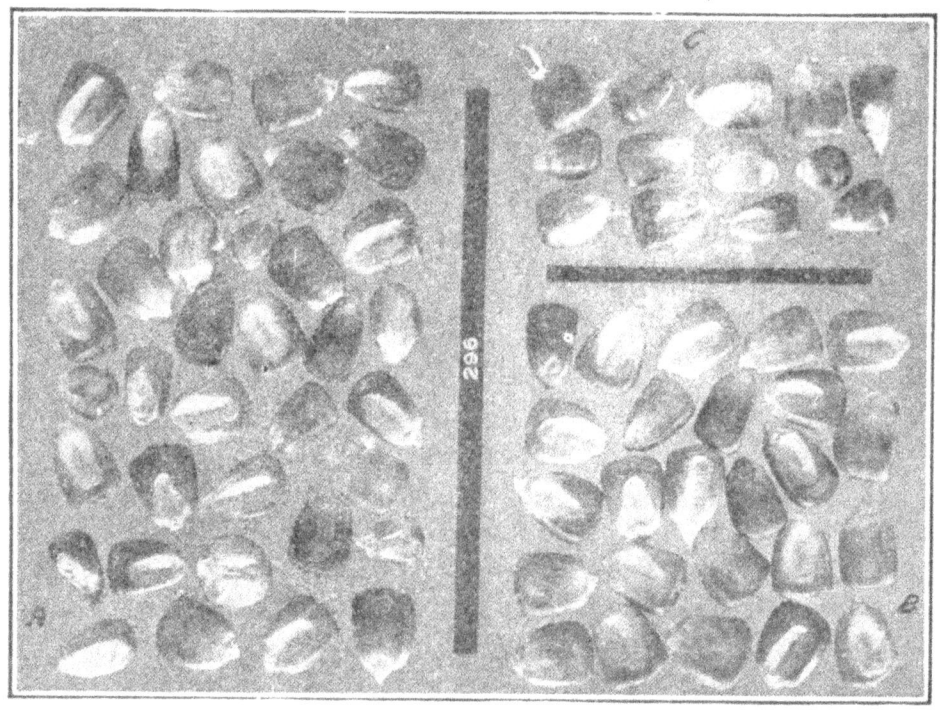

Figure 28.—(A) Corn as shelled from whole cob. (B) Corn after it has been graded. Note uniformity of kernels. (C) Small and irregular kernels removed by the grader.

middle of the ear, owing to imperfect growth. Examine an ear of corn and you will see the irregular kernels at the tip and butt and a few in the middle. Irregular kernels are easily removed from any sample of corn by running it through any of the modern corn graders.

Corn graders are fitted with screens with holes of certain sizes and shapes. As the corn is run through, all the small and irregular kernels are separated out, leaving only the regular ones and the kind a planter can plant uniformly. A corn grader will cost about $10.00, and is well worth the money to any one who is planting any considerable amount. If one has no corn grader it is advisable to shell off tip and butt kernels and keep them out of the seed corn.

Planting.—Corn is, as a rule, planted in the spring at about the time danger of frost is past. There is, however, no set date for planting, but a good practice to follow is to plant as early in the spring as the soil can be and is well prepared and is warm enough so that the seed will germinate. There is nothing gained by planting corn in cold, wet, poorly prepared soil; for, in such condition, seed, even if good, is very liable to rot in the ground. Probably one very general mistake is made in planting corn too deep. Corn cannot be made to root deep simply by planting deep. The roots will go wherever the soil conditions are most favorable. On ordinary land it is not wise to plant corn more than from one to two inches deep.

Figure 29.—Planting Corn. Straight rows indicate good workmanship and are a joy throughout the year.

Questions:
1. What is a good crop to precede the corn crop?
2. Why prefer fall plowing to spring plowing for corn?
3. What is gained by grading seed corn?

Arithmetic:
1. It costs 10c. per bushel to grade seed corn. What does it cost per acre if a bushel of corn plants 7 acres?
2. If a man were to spend five hours shelling off tip and butt kernels and picking out irregular kernels from a bushel of seed corn, how much would it cost him, if his time is worth 14c. per hour? How much would it cost him per acre? (See Example 1.)

REASONS FOR THE CULTIVATION OF CORN

Conservation of Moisture.—One of the main reasons for cultivating corn is to save moisture in the soil. There are two ways by which moisture is taken from the soil: first, by the growing crops and, second, by evaporation from the surface of the soil. The water that is evaporated is worse than wasted. It is lost to the crop and its evaporation tends to make the soil cooler. (Demonstrate this by moistening the hand and exposing it to the air.) The hot sun, the moving air, and the wind, greatly hasten evaporation; hence it is evident that, if the part of the soil in which the moisture is held can be separated from the sun and wind, much less moisture will be lost by evaporation.

Moisture moves in the soil by capillary force. For example, two particles of sand lie close together and one is wetter than the other. The dryer one will attract some of the water from the other until both are equally wet.

Experiment.—This capillary movement of water may be seen very plainly by following this plan as suggested: Fill a glass half full of sand and pour in enough water to thoroughly wet the sand, being careful not to wet the sides of the glass. Now fill the glass with dry fine sand and watch the moisture creep upward. The water is moving by capillary attraction. Moisture will move by capillary force in any direction, but always from moist to drier soil.

Surface Mulch.—If you will go out in a field, on a bright day, where cultivation is being done, you will notice that very soon after the cultivator has gone over the ground the surface soil begins to look dry, while the surface soil where it has not been cultivated appears more moist. Moisture is being evaporated from both surfaces, but where the soil is firm moisture moves up from below to replace that evaporated; and this process will continue if not checked, until the soil is robbed of all available moisture. The cultivated portion is so loose that the moisture cannot readily pass up through it, as the particles of soil are not close together, hence evaporation is checked.

Too much attention can hardly be given to maintaining this surface mulch, or loose soil on top. A shower of rain just sufficient to pack the surface may be more injurious than beneficial to a crop, by causing a great loss of moisture, unless the soil is cultivated soon after the shower.

Destruction of Weeds.—Another reason for cultivating

CULTIVATED CROPS 65

Figure 30.—A two-row cultivator at work. Many farmers are now using these larger machines. One man can do nearly twice as much work with such a machine as he can with a one-row cultivator.

is to destroy weeds. The smaller the weeds are the easier it is to kill them. On this account it is important that soil be so worked that most of the weeds are destroyed before the corn is planted, as it is difficult to cultivate very young corn and the weeds may get ahead of the corn. It is also cheaper to cultivate with a large harrow than with a cultivator. Of course corn may be harrowed after it is planted or even after it is up, but the less harrowing necessary at this time the better, as more or less corn is injured every time the field is harrowed.

In fields well prepared before corn is planted, very few weeds will be seen, and those appearing are small and easily killed at the first cultivation.

Other Objects.—Cultivation aerates the soil, i. e., opens it and lets air into it, thereby stimulating the decomposition of vegetable matter and liberating more plant food.

Cultivation also keeps the soil loose, so that rain which falls may be readily absorbed, thus making a larger supply upon which to draw when a dry season comes.

Questions:
1. What is the main reason for cultivating corn?
2. In what two ways is moisture taken from the soil?
3. How does cultivation check evaporation?
4. Give another reason for cultivating corn.

Arithmetic:

1. A team travels 2¼ miles to cultivate an acre of corn planted 3⅔ ft. apart each way. How far must it travel to cultivate 8 acres?
2. If a team travels 18 miles with a harrow 12 ft. wide, how much land would it harrow?
3. How many times can one afford to harrow land to save one cultivation? (See two examples above.)

METHODS OF CULTIVATING CORN

Depth to Cultivate.—There seems to be an unsettled question among farmers as to how deep to cultivate corn. Apparently there is no rule that one can safely follow, for the conditions vary with soils and seasons, so that it is largely a matter that must be settled by the individual farmer and depends entirely upon the depth of the corn roots. In a dry loose soil corn roots will grow nearly straight down, while in a heavy or more moist soil they will spread out near the surface of the ground. Roots naturally grow where there is available plant food; and that, we have learned, is where there is heat, air and moisture. In wet years they find this condition near the surface, and in dry years or in dry soil they must go deeper down for the plant food. The accompanying cut shows how corn roots usually grow. (Figure 31.)

Roots of Corn.—A very interesting study of the root system of corn may be made by taking a rather blunt wooden paddle and carefully scraping away the loose soil between two hills of corn, until the roots are exposed. One may then observe the roots, how far they spread out from the hill of corn and how near the surface they grow. As a rule, when the corn is a foot high the roots from the rows will be overlapping and within one to four inches from the surface, depending upon how wet the soil is and how recently and how deeply cultivated.

Results of Deep Cultivation.—From the above facts it is quite evident that, if the cultivator is run too deep, some of the roots will be cut off. The roots are the feeders of the plants; consequently every one that is cut off decreases the amount of moisture and plant food the plant will get. The effects of too deep cultivation may be seen by cutting down in the soil four inches, with a sharp spade, two to four inches from a hill of corn. Then note results. If it

CULTIVATED CROPS 67

is a dry, hot day the leaves will soon begin to curl up on the plant thus injured, showing that a portion of its water supply has been cut off.

It is necessary, however, to cultivate to kill weeds, to let air into the soil and to form a surface mulch to save moisture; and many times it is necessary to cultivate deep enough to injure corn roots in order to accomplish these various things; but the aim should always be to cultivate no deeper than necessary. If deep cultivation is to be practiced at all it should be done while the corn is small, as it is injured less at this time.

Cultivator.—The kind of cultivator used has much to do with the depth of cultivation. If a cultivator with two large shovels on a side is used, it must be run deeper to cover all the space between the rows, than one which has three, four or five shovels on a side. The small shovels and more of them do finer, shallower work than the large shovels; but where corn has been neglected until the weeds are large, the larger shovels are better, because they do not clog up so easily and because they plow out the weeds instead of cultivating them.

Surface Cultivators.—At present many farmers are using what are called surface cultivators. In place of shovels there are two or more knives or blades that run an inch or so below the surface of the ground, separating the surface soil from the soil below and cutting off just

Figure 31.—The root system of corn. Kansas Bulletin No. 147

below the surface all weeds growing between the rows. If possible, examine cultivators with these different kinds of shovels and note the work they do.

Check vs. Drills.—Many farmers drill in their corn, i. e., plant it in rows only one way. It can then be cultivated only one way and the weeds growing between the hills cannot be reached with the cultivator. If these weeds cannot be covered by having the cultivator throw earth against the rows, they must be pulled by hand or let grow.

Other farmers plant their corn in check rows. As they can then cultivate it both ways, they can get all the weeds with the cultivator, excepting an occasional one growing in the hills. Try to look over fields of corn planted each of these ways, at different times during the summer, and see which fields are the cleaner. If you can find corn planted each way in the same field or on the same farm, and receiving the same number of cultivations, it will be a better comparison. As one of the main objects in growing corn is to clean the land of weeds, it is better on weedy land to plant corn so that it can be cultivated both ways. If corn is cultivated both ways, it is easier to keep the surface smooth and level, a condition which is desirable, as a ridged surface is hard to work down, and more surface is exposed, causing more evaporation.

Questions:
1. Why is it not wise to cultivate corn too deeply?
2. What can you say about different types of cultivators?
3. What is said of planting corn in drills or in check rows?

Arithmetic:
1. If the time of a man and team is worth $4.00 per day, what is the cost per acre to cultivate, if they cultivate 8 acres per day? How much does it cost to cultivate an acre of corn six times?
2. How many bushels of corn at 54c per bushel must a farmer get to pay for cultivating his corn six times?
3. If the time of a man and three horses is worth $5.00. What is the cost per acre, if they cultivate 14 acres per day? (Three horses can draw a two row cultivator.)

SELECTION OF SEED CORN

Selection Neglected.—A comparatively small amount of seed corn is needed each year, on the average farm, as one bushel will plant from six to eight acres. On this account the matter of saving seed corn is likely to be neglected, as farmers are very busy in the fall with other duties. Were the saving of the price of seed the only advantage gained in selecting seed corn on the home farm, one might be justified in neglecting it; but this is by no means the case.

Adapted to Localities.—Corn in some respects is a tender plant, very easily affected, unfavorably by cold weather conditions or cold wet soil, and favorably by warm weather and warm soil. On this account corn grown under one condition for several years becomes adapted to those conditions and is not well suited to other conditions. No locality is suited to produce seed corn for any very large territory. Corn that does well in the north will grow and do well farther south, but as a rule a larger corn can be produced on most of the well drained soils of the south, and will yield more than the comparatively small corn grown farther north. Corn suited to Indiana conditions will grow if planted in northern Minnesota, but in average years it will not mature well, as the season is too short. The same varied conditions may be found on different farms in the same locality. Farms with a light warm soil, or well-drained farms on which the soil is kept highly productive by good methods of cropping, manuring, etc., can grow and mature a larger variety of corn than farms in the same neighborhood with heavy, and poorly drained soil or soils in poor condition.

Figure 32.—Two varieties of Dent Corn growing side by side and given similar treatment That on the right is mature, as shown by the drooping ears; that on the left is still quite green, as shown by the erect ears. Both are Yellow Dent Corn, but one later than the other by being grown under different conditions.

If a person wishes to get the best results from growing corn, he cannot afford to neglect selecting his own seed from his own farm. By selecting the best ears of corn from the best stalks one gets seed from the plants that are best adapted to the conditions of the farm, as shown by their superior development the previous year.

Large Varieties.—A mistake very commonly made is to select too large or too late a variety of corn. Every one likes to grow large ears of corn, and on this account, when seed is secured from some other community or from seedsmen, a larger variety than is suited to the conditions is likely to be choosen. Large ears of corn are not necessary to large yields, and it is far better to be sure of a good crop, by using a variety that will mature, than to attempt to grow too large a variety and have a partial or complete failure occasionally.

Varieties May Be Made Larger.—It is well to select a variety of corn that will be quite sure to mature in your locality. If the soil is well drained, well cultivated and kept at a high state of productivity by growing clover occasionally and by keeping live stock and manuring it, and if the climate will permit the growing of a larger variety, one can in a few years make the variety larger by selecting the larger ears. Such conditions will practically insure a good crop of corn each year, unless one selects too large ears ears and thus makes his variety too late. If in a few years you cannot improve the corn to the size you wish it, it is likely that your conditions are not favorable for a larger variety; and, were you to get a larger variety from some other locality, you would be very likely to lose your crop or have soft corn, in the ordinary years.

To Make a Variety Early.—If one wishes to make a variety of corn one is growing earlier, one can do so by selecting the ears that ripen first. Such a selection cannot be made after all the corn is ripened. If one can not select the seed when the corn is ripening one can make some progress toward earliness by selecting the small ears of corn with comparatively shallow kernels. Large ears with deep kernels are very seldom found in an early variety of corn.

Questions:

1. Give two reasons why it is advisable for a farmer to select his own seed corn from his own farm.
2. Why is it better to have a variety of corn that is a little too small rather than one that is too large?
3. Give two ways by which a variety of corn may be made larger.
4. To get corn that will ripen earlier, how and when would you select it in the field? How select it from a large number of husked ears?

Arithmetic:

1. A plants 7 acres of corn with 1 bu. of seed and it yields 40 bus. per acre. How many bushels of corn does he get? Extra good seed would have increased the yield 20%. How many more bushels of corn would he have received had he used good seed? How much would the increased yield be worth at 54c. per bushel? How much would a bushel of extra good seed corn have been worth to that farmer?
2. There are 3,240 hills of corn on an acre when planted 44 inches apart each way. If a farmer gets 3 10-oz. ears from each hill, how many bushels of corn will he produce?

HOW TO SELECT SEED CORN

Kind to Select.—If one is to get the best seed ears from a field of corn, one must have well in mind what a really good ear of corn looks like, and select only such ears. A great advantage of selecting seed corn in the field over selecting it from a load of husked corn is that the stalks may be considered as well as the ears. No matter how good an ear of corn may be, it should not be taken from a poor plant. Usually good ears come from good plants, but there are exceptions. It is well to select more seed corn than is needed. Then another and more careful selection may be made in the spring before planting.

Time to Select Seed Corn.—In order that seed corn may be sure to keep over winter and still germinate readily it must be taken from the husk and placed where it can dry out before freezing weather. If one weighs an ear of freshly husked ripe corn, then leaves it in a living room for a month and weighs it again, it will be found that it has lost in weight. The loss in weight is from the evaporation of moisture. Moisture is detrimental to seed corn. Select and husk seed corn before there is danger of a killing frost so that the seed will not be injured by frost and so it will have time to dry out before freezing weather.

Condition.—The first thing to consider in an ear of corn for seed is condition. It must be firm and solid to the touch and heavy, not light and chaffy. Loose or soft kernels indicate immature ears, which must be avoided, as corn from such ears is not likely to germinate and, if the kernels do germinate, the plants are likely to be weak. The kernels should be bright in color and free from mold or injury.

Shape of Ear.—Ears should be uniform in shape and size, and each ear should be as nearly the same size at tip and butt as possible. The tips should be well filled out, as this indicates hardiness and well matured corn. Large butts should be avoided as they indicate coarseness and are hard to dry out. There are, also, more irregular kernels on these large butts than on properly formed butts.

Size of Ear.—The size of ears will depend upon the variety and the locality. But do not select too large ears, as they will have a tendency to make the variety later, which may result in considerable loss in unfavorable seasons. Select the medium sized, well matured ears as nearly uniform in size as possible.

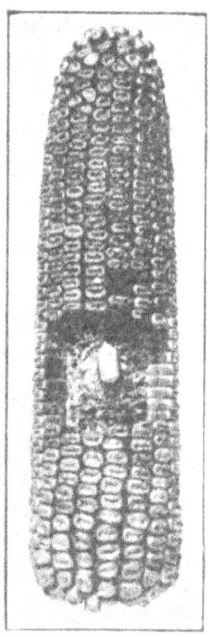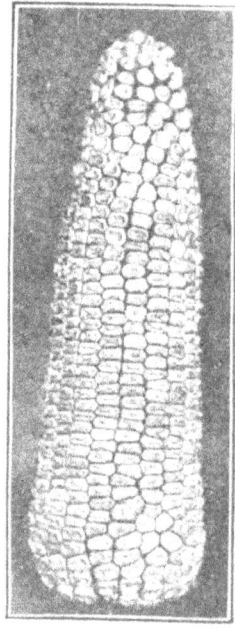

Figure 33.—The ear on the left represents a good type to select for seed. It has even, regular rows and kernels and a good proportion of corn to cob. The ear on the right is just opposite in character, and is undesirable.

Kernels.—Ears with kernels as nearly uniform in type as possible should be selected. There are good ears of corn with different types of kernels, but for any one variety it is important that the kernels be uniform, as only such kernels can be planted uniformly by machinery. The most desirable kernels are deep, indicating a large amount of corn in proportion to cob, but the point can be overdone, as deep kerneled varieties are usually late in ripening.

Space between Kernels. —It is desirable to have just as much corn around the cob as possible; consequently any space between the kernels is to be avoided. These spaces are caused by rounded kernels, and are more common and larger in flint than in dent varieties.

Selecting.—It is a comparatively small task to go through the field with a sack and select the desired ears, or when the corn is husked from the standing stalks the ears may be selected as the husking is being done and the seed ears thrown into a sack or small box on the side of the wagon. This manner is preferable to selecting the best ears from the load or crib after it is husked, as the stalks may be considered in the selection.

Application.—It is hoped that the boys and girls who read this lesson will take an interest in the selection of seed corn, and at least observe how it is done on the home farm and on other farms. Some of the spare time during the fall may be profitably spent in helping with the selection of seed, and much information gained by asking your fathers and brothers why they select certain ears and discard others. Good seed corn is worth from $2.00 to $5.00 per bushel, and, if some boy wishes to make a little spending money, he can do it quite easily by carefully selecting and storing a few bushels of seed corn. Farmers will gladly pay a good price for such seed.

Questions:
1. When should seed corn be selected, and why?
2. What are some of the advantages of selecting seed corn in the field from the standing stalk over selecting from the load or crib?
3. For what reasons would you select ears that are sound, with kernels tight on the cob?

Arithmetic:
1. A bushel of seed corn is worth $3.00 and will plant six acres. What is the cost of seed per acre?
2. A fair sized ear of corn will weigh about 10 ozs. What part of a pound does it weigh? How many such ears of corn in a bushel? (A bushel of ear corn weighs 72 lbs.)
3. A boy selects 200 10-oz. ears of seed corn in one day. How many bushels does he save? How much is it worth at $3.00 per bushel?

STORING SEED CORN

To Keep Germ Uninjured.—We have learned (page 57) that every kernel of corn contains an embyro or germ, which is a very small, live corn plant. If the kernel is to be of any value as seed, this germ must be kept alive and strong.

This little plantlet, or germ, is very similar to any plant. Freezing, under certain conditions of moisture, will kill it. This germ can stand freezing only when quite dry, as when in this condition it is dormant. Trees and other plants that live from year to year are very liable to be killed by cold winter weather, if kept growing until late in the fall. Under normal conditions such plants stop growing several weeks before cold weather sets in; which gives them a chance to "harden up" or, as we might say, "ripen." It is evident, then, that, if we would keep our seed corn in good condition, it must be so handled as to prevent injury to the germ in each kernel.

Keep Dry.—The first essential is to select the seed ears before they have a chance to freeze in the field, for many times the corn may not become sufficiently ripe to be thoroughly dry; and if not dry, freezing injures the germ. After the husk has been removed, the ear will dry out rapidly, if placed where it has an opportunity to do so. Seedsmen appreciate the necessity of drying seed corn immediately, and they store it in a room in such a way that air can circulate about it freely and thus carry off the moisture. They very often use artificial heat, as stove or furnace heat, to assist in this drying operation.

In the Attic.—A farmer, as a rule, saves only a small amount of corn and cannot afford a special storehouse for it. Probably the most satisfactory way of drying corn and keeping it dry, on the farm, is to store it in the attic over the kitchen. Here ventilation can be supplied by opening windows, and the heat from the kitchen stove assists in drying out the corn and in keeping it dry. A basement in which there is a furnace, so that the corn will be kept dry and so that there will be good ventilation is also a very good place for seed corn. Where one does not have these facilities or has more corn than one can store in an attic or dry basement satisfactorily, it may be placed on a barn floor or in a vacant room in the house or other building. It should not be piled over eight to ten inches deep, as it may heat or sweat, if piled deeper. Good circulation of air should be supplied, as this aids in drying the corn, and it is very essential that it be thoroughly dried before cold weather. If corn is thoroughly dried and kept dry it will stand freezing; but it is much better if it can be kept where the temperature is slightly above freezing.

Seed corn should never be placed above a stable in which animals are kept, or over a bin of grain, as the steam and breath from the animals, or the steam that may rise from a bin of grain, if it heats even slightly, will keep corn moist enough to greatly reduce the vitality of the germ.

Good Seed Essential.—A kernel of corn is a very little thing, but it is a very important factor in the production of good yields. Very little time is necessary to select and care for all the seed corn needed on the average farm and few farmers can afford to neglect this part of the business.

Neglecting to save and properly care for seed corn may save one or two days' time in the fall, but it may also mean that poor seed corn or corn not well adapted to one's conditions will have to be planted the next year. Poor seed corn means a partial or total loss of the corn crop, which may result in a very great financial loss.

Questions:
1. What will injure the germ in a kernel of corn?
2. Why should one take seed corn in from the field before frost?
3. How do seedsmen store corn, and why?
4. How may farmers store their seed corn?

Arithmetic:
1. If it requires 20 ears of corn to plant an acre, how many ears are required to plant 40 acres?
2. If a man can select 800 ears of corn in 2 days, how much will it cost him to gather the 800 ears, if his time is worth $2.00 per day?
3. How many bushels of corn in 800 ears of corn weighing 10 oz. each (72 lbs. per bushel)? How much is it worth at $3.00 per bushel?

METHODS OF STORING SEED CORN

Drying.—Free circulation of air about seed corn is necessary to dry it out, consequently many devices have been used for storing it easily, quickly and in such a way that this end will be accomplished.

The old practice, of braiding several ears together by the husks and hanging them up, is a satisfactory way to keep the corn, but requires a great deal of unnecessary labor.

A Simple Device.—A very simple and practical device for putting up seed corn is illustrated in Figure 34. This device is called a "corn tree." Any boy who can use a saw and hammer can make one in a short time. To make it, saw a 2 x 4, or better, a 4 x 4 off five or six feet long. To the bottom end of this spike a plank about 12 inches square, to form a base sufficiently large so the tree will stand firmly erect on the floor. It is well to put some short braces from the edges of the plank up to the 4 x 4 to stiffen it. A row of finishing nails, nails with small heads, are driven in each side, of the 4 x 4 and about 2½ inches apart. An ear of corn is easily stuck on each nail by jamming it on butt first. The nail sticks into the pith of the corn cob. This tree may be placed in the attic or any other convenient place where the corn will be kept dry. If the tree is six feet high, it will hold about 100 ears, or enough to plant about five acres. If one wishes to put more corn on the tree, the corners of the 4 x 4 may be beveled off, making it eight-sided. There will then be room for eight rows of corn. Thus a tree six feet high will hold 200 ears. It is well to plane the 4 x 4 smooth, so that numbers may be placed at the base of each nail, thus making it easy to number the ears, if one wishes to test each ear for germination.

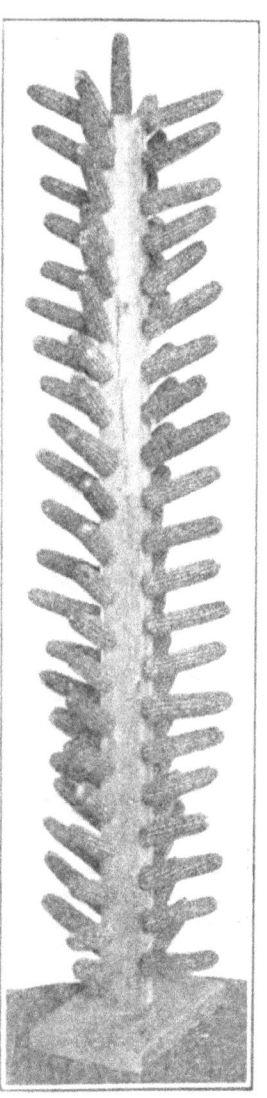

Figure 34.—A simple device for putting up seed corn to dry.

The double string method is likewise a very practical means of putting up seed corn. Take a piece of binding twine about fourteen feet long. Tie the two ends together. Then string up the corn as indicated in Figure 35. The strings with from ten to fourteen ears of corn in each are easily handled and may be hung from the rafters or other convenient places.

CULTIVATED CROPS

Figure 35. Showing the double string method of storing seed corn.

Slatted Shelves.—Strips of timber 1x4 or 2 x 4 stood on end with lath nailed on either side to form shelves make a very good means of putting up seed corn. This method is often used by seedsmen and permits of storing in good condition a large amount of seed corn in a comparatively small space.

Patented Corn Hangers.—There are numerous patented seed corn hangers on the market. Most of them are good and afford a very satisfactory means of putting up seed corn. The only objection to them is the cost. Very good wire seed corn hangers may be made from electric weld woven wire fencing. Your state Experiment Station will furnish you more complete information about making any of these seed corn hangers.

Wire Baskets.—Long, slender wire baskets are very easily made of poultry netting or other closely woven fencing. The two ends of a strip of fencing three to five feet long are fastened together, and a board or another piece of the netting put in for the bottom. Such a basket will hold considerable corn, and hold it in such a way as to allow it to dry readily. These baskets are preferable to the other methods given above only when a large amount of seed corn is to be stored.

Questions:
1. Why is a good circulation of air about seed corn necessary?
2. Describe a corn tree.
3. What can you say of wire baskets for storing seed corn?

Arithmetic:

1. How many feet of lumber in a piece of 4 x 4 6 ft. long? How much is it worth at $30 per thousand feet.

2. If it takes about 2 hours to make a corn tree, how much does it cost for labor, it the boy's time is worth 6c. per hour?

3. If it takes 14 ft. of twine to hang up 10 ears of corn, by the double string method, how many feet will it take to hang up 200 ears of corn? How much will it cost, if twine is worth 8c per pound? (500 feet per pound.)

CORN FOR SILAGE

Corn silage is being used throughout the Corn Belt in ever increasing amounts. This is due to the fact that corn is the surest crop that is grown in that section of the country. The corn crop is more nearly controlled by the farmer, and less likely to be injured by drouth, hot wind, insect pests, diseases and hail than such crops as grass or grain. This advantage is due to the fact that the corn crop is cultivated and may be kept growing well when the dry weather or hot winds destroy or greatly reduce the hay or grain crop.

Live stock can not be profitably kept without an assured supply of suitable feed. Where silage is not available it is often necessary to sell off a large part of the stock in dry years, because the pastures are short or because not enough hay was produced to winter them. With a silo full of good corn silage one can be reasonably sure of feed for stock either winter or summer.

Feeding Value of Silage.—In silage all the nutrients produced in the corn crop, both in the ear and in the stock, are saved. Dry corn stalks make fairly good feed, but usually half of the stalks are left uneaten by the stock. These stalks are not only wasted but are troublesome in the yard or barn. Good silage is much more palatable and relished more by cattle than dry stalks. Not only are more of the stalks eaten in the form of silage, but, because the silage is relished better, a larger part of that eaten is digested. There is no way known by which a large part of the whole corn plant can be made more palatable for live stock than in silage. Silage, because it is succulent (juicy), comes more nearly furnishing summer conditions for stock in the winter than any other farm feed.

Corn for Silage.—When silos were first used corn was put into them quite green. It was found that this silage was very sour, and sometimes animals did not like it, or, if they did eat much of it, it did not agree with them. It has been found, since silage came into more common use, that corn must be really ripe or mature when it is cut and put into the silo, if first-class silage is to result. Now farmers usually grow for silage just the same variety of corn that

Figure 36.—A good stand of corn for silage.

they grow for ears. They plant it at about the same time in the spring, and cultivate it in the usual way. Corn for silage is generally planted about 50 per cent more thickly than for ears. It is more commonly planted in drills than in check rows. It may be planted in check rows, if desired.

Time to Cut.—The most important thing about getting good silage is the time of cutting the corn. There are no reliable rules to go by, because in wet years conditions are quite different than in dry years. In wet years the ears of corn may be entirely ripe, while the stalks and leaves are still green. In dry years the stalks and leaves may be quite dry before the ears mature. A crop of corn increases in feeding value up until the time it is mature. It is, therefore, desirable to have the corn mature when cut for silage.

but it must be cut while the stalks are still green enough and have enough sap in them so the silage will pack down well in the silo. The ideal condition in which to cut corn for the silo is when the ears are nicely ripe and the stalks and leaves are still green.

Cutting Silage.—Corn for silage is usually cut in the field with a corn binder, loaded at once on wagons and hauled to the silage cutter. Here it is cut into short lengths, from ⅜ to ¾ inches in length, and elevated into the silo. One or two men are kept in the silo while it is being filled, to keep the silage well packed. It is important to pack the silage thoroughly so as to crowd out as much air as possible. Air in silage causes it to spoil. Silage is kept in a silo because the silo keeps the air out. The machinery for cutting silage is quite expensive, so it is advisable, whenever possible, for several farmers to co-operate in buying a silage cutting outfit. One can fill six or more silos in one year.

Questions:
1. Why is silage and especially corn silage important on live stock farms?
2. By what means may the largest proportion of the corn crop be saved for feed? Why?
3. Describe the method of growing corn for silage, and when it should be cut.

Arithmetic:
1. If it costs $20 per acre to grow and store silage, how much does it cost per ton, if there is a yield of 9 tons per acre?
2. Compared with bran at $25 per ton, silage is worth $3.75 in feeding value. How much is an acre of silage yielding 9 tons worth?
3. If a silage-cutting outfit can cut 75 tons of silage per day, how many days would be required to fill six 100-ton silos?

THE POTATO CROP

Importance.—Since potatoes are a side issue on many farms they are often grown without receiving the care necessary to insure a successful crop. A great deal of work is required to grow an acre of potatoes, hence the importance of fitting the soil and caring for the crop, so that a good yield may be expected. A fair crop of potatoes is worth $40 per acre. A fair crop of grain is worth $10.00 per acre. Care in preparing the soil so as to increase the yield 10 per cent means an increase in value of $4.00 in the potato crop and but $1.00 in the grain crop. Thus, when a crop that brings a comparatively large income per acre is raised, one can afford to put more expense on fertilizing or preparing the soil or on other operations, as cultivating, etc., than when crops yielding less in money value are grown.

Seed.—About ten bushels of seed potatoes are required to plant an acre. The best seed potatoes are secured by selecting them from hills in which there are large numbers of uniform and desirable potatoes, rather than from hills with some large and some small ones. See Figure 39. Of course such selection cannot be made in the spring. So, if one did not make the selection in the fall at digging time,

Figure 37.—A potato field. Note weedless, straight rows.

the next best thing is to select good, smooth, uniform, shallow-eyed potatoes from the stock at hand.

It is not wise to plant small potatoes, for it is unreasonable to expect to raise better potatoes than are planted. Very little ill effect will be realized from using small potatoes for a year or so, but if one continues the practice one can but expect to raise small potatoes.

Prevent Seed from Sprouting.—Potatoes are likely to begin to sprout as soon as the weather gets warm. This sprouting is undesirable, as the sprouts take nourishment which should be saved to nourish the young plant when started in the field. Keep the seed in as cool a place as possible without freezing it, and where it is dry. It is a good plan to keep seed potatoes in baskets or slatted boxes piled up in a cool cellar so that the air can circulate freely about them.

Scab.—The rough blotches on the surface of potatoes are called scab. The disease is caused by certain spores or seeds, just the same as diphtheria or other contagious diseases are caused by germs. To prevent scab the spores of the disease must be destroyed. The spores may live over winter in the soil on which scabby potatoes were grown the year before. They may get into the soil with manure from animals that have been fed scabby potatoes, or they may

Figure 38.—A manure spreader

be on the seed planted. The latter is the most common way of spreading the disease, and, as the seed is very easily treated to prevent scab, there is very little excuse for getting scabby potatoes in this way. Formaline is a liquid which may be purchased at any drug store for from twenty-five to fifty cents per pint. A pint mixed with thirty-five gallons of water makes a solution which will destroy the scab spores on seed potatoes, if they are soaked in it for two hours. Treat for scab before cutting the potatoes. If cut first, some of the pieces may stick together and the spores in between will not be reached by the solution.

There are several other common diseases affecting potatoes. These are quite easily controlled, if one is familiar with them. Every potato grower should be thoroughly informed regarding potato diseases and their control. Write to your experiment station for full information.

Cutting Seed Potatoes.—Experiments have proved that rather good sized pieces give larger yields than small pieces. It is well to cut the pieces to about 1 oz. in size, being sure to get at least one good eye in each piece. The large pieces furnish more food for the plants until they get their roots started than do the smaller pieces. Sometimes whole halves are planted, and often pieces having at least two eyes.

Questions:
1. Why can a farmer afford to spend more time preparing an acre of soil for potatoes than for grain?
2. How should seed potatoes be kept during the winter?
3. How does sprouting injure seed potatoes?
4. How is the disease known as scab spread? How is it treated?

Arithmetic:
1. If potatoes are planted in rows 36 in. apart, with hills 16 in. apart in the row, how many sq. ft. of space will each hill occupy? How many hills will there be on an acre? (There are 43,560 sq. ft. in an acre.)
2. If there are 10,890 hills of potatoes on an acre and one 1 oz. piece is planted in each hill, how many bushels of seed will be required to plant an acre?
3. If there are 10,890 hills of potatoes on an acre, how many pounds must each hill yield to produce 300 bushels per acre?

PLANTING AND CULTIVATING POTATOES

The Soil and Its Preparation.—Potatoes require rich, moist, mellow soil; and, as the tubers must grow under ground to protect them from the sun, it is well to have the soil mellow to quite a depth, six or eight inches. It is evident that the land must be plowed to a good depth. Fall plowing is preferable, as it gives the soil a chance to become firm and settled and be acted upon by the weather. Spring plowing, unless very thoughly disked and harrowed, is likely to be too loose and to dry out. It is also more likely to be lumpy.

Clover sod, or land that has grown clover the previous year, and was plowed in the fall, is the ideal soil for potatoes, especially if the land was top-dressed with manure before it was plowed. A very excellent way to prepare such land is to top-dress it with good stable manure in the fall on the clover sod before it is plowed. Then disk it to cut up the sod and mix the manure with it. Then plow in the fall. This treatment gets the clover sod and the manure well pulverized and mixed together and turned under where the tubers are to grow. It insures them a rich, mellow place. If soil is very light and sandy, it would be better to plow the land in the fall without manuring it. Manure it during the winter or spring and disk the manure in, thus keeping it near the surface.

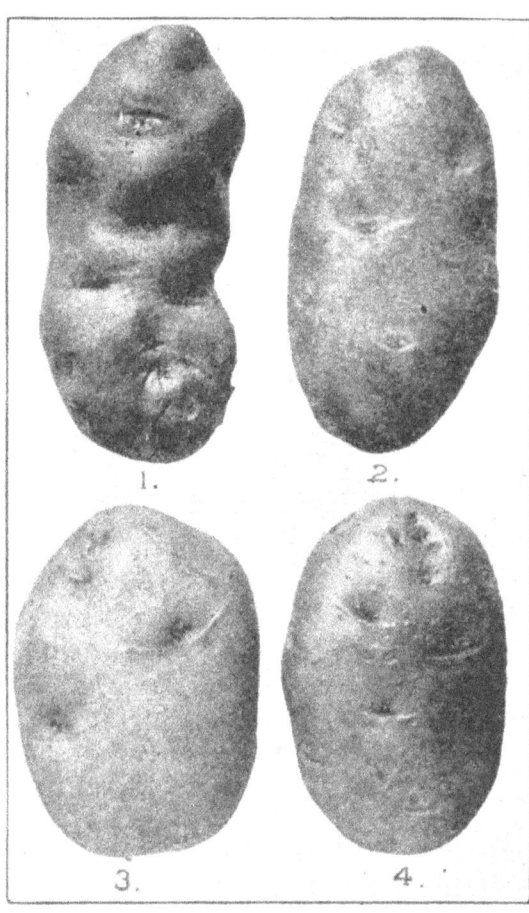

Figure 39.—Types of potatoes. 1 is a rough, deep-eyed type, not desirable for any purpose. 2. A good type of Burbank. 3. A good type of Carmen No. 3. 4. A good type of early Ohio. Note smooth surface and shallow eyes of Nos. 2, 3 and 4.

If land has been treated as suggested above, disking and harrowing it a few times in the spring will put it in excellent condition for planting.

If the land has not been prepared in the fall, then the same manuring and disking should be done before the land is plowed. Then plow, harrow and disk until the soil is well packed down. It is important that spring plowing be well harrowed to assist in firming it, that it may not be so loose as to dry out too quickly.

Planting.—Potatoes may be planted from early spring to early summer. Early planting is usually preferable so that they will have a chance to make their growth before the dry, hot weather comes. Potatoes grow better in rather cool weather when the soil is reasonably moist. They are usually planted in rows about 36 inches apart and the pieces are dropped from 14 to 18 inches apart in the row. About 4 inches is a good depth to plant them. If one has a horse planter it is a very easy matter to plant potatoes. There are also satisfactory hand planters. If but one half an acre to an acre is raised, as is the case on most farms, they may be easily and well planted by marking the land with a corn marker, then plowing a furrow, for each row, with a common walking plow or a shovel plow, dropping the seed in these furrows by hand, then covering with a plow or by harrowing crosswise.

Blind Cultivation.—If potatoes are planted with a hand planter or by dropping into furrows as suggested above, it is a good plan to go into the field with a cultivator after they have been planted a few days and give the plot a good cultivation. This is called blind cultivation. Set the shovels so as to throw the dirt on the row, thus making a ridge over each row. Follow this every few days by harrowing.

Throwing the dirt in a ridge over the rows, then later leveling it off with the harrow, keeps the soil mellow over the rows and prevents weeds from starting, making it much easier to keep the rows clean.

Cultivation of potatoes should continue at frequent intervals, from the time they are planted until the vines cover the ground, to keep down weeds and to check the evaporation of moisture. Potatoes require a great deal of moisture, and a lack of moisture at any time reduces the yield. As in cultivating corn, care must be taken not to cultivate deep enough to injure the roots. Except when the soil is cold and wet, level cultivation is preferable to hilling.

Spraying.—It is now quite impossible to grow potatoes successfully without spraying the vines one or more times, While they are growing. They must be sprayed both for bugs and diseases. A spray containing Paris green or other poison to kill the bugs and Bordeaux mixture or other fungicide to check the blight is commonly used. This saves time, as the combined spray is as easily applied as a single purpose spray, and is quite as effective. for full particulars write to your State Experiment Station.

Questions:
1. For what reasons would you prefer fall plowing for potatoes?
2. Describe a good method of preparing clover sod for a crop of potatoes?
3. How are potatoes planted? How cultivated?

Arithmetic:
1. It costs 35c. per acre to disk land. A farmer disks his potato field twice after manuring and before plowing. What must be the increased yield to pay for the extra work of disking twice, if potatoes are worth 35c. per bushel?
2. It costs 50c. per acre more to plow 6 in. deep than to plow 4 in. deep. How much does one get for his extra labor, if land plowed 6 in. deep yields 10 bus. more than land plowed only 4 in. deep, if potatoes are worth 35c. per bu.?
3. If it costs 50c. per acre to cultivate potatoes, how much must each cultivation increase the yield to pay for the cultivation, if potatoes are worth 35c. per bushel?

ROOT CROPS

Importance.—Root crops, such as mangels, rutabagas, turnips, stock carrots and sugar beets, are grown quite generally as feed for stock. They are especially important on farms not supplied with ensilage. On farms where stock equivalent to ten cows or less is kept, it is seldom practical to use silage. Good stock feeding requires that some sort of succulent feed be provided. Root crops may be grown with the machinery ordinarily found on the farm, and do not require an expensive building for storage. For these reasons it is decidedly practical to grow root crops on farms keeping only a small amount of stock, if the stock is an important factor on the farm.

Other Uses.—Sugar beets are grown much more extensively for the manufacture of sugar than for stock feed. Rutabagas and turnips are grown quite extensively as vegetables for human food.

Culture.—Root crops require a great deal of hand labor. On this account it is important that they be planted on very rich land that will produce a heavy yield. Ten tons per acre is a common yield. It is more practical to have the soil so rich by manuring it, and in such fine condition by thorough cultivation, that yields of from 20 to 40 tons may be secured. Deep plowing, preferably in the fall, to the depth of from 8 to 12 inches is desirable for roots. This should be followed by very thorough disking the following spring. The land may and should be disked several times before the roots are planted. Soon after cornplanting is best time to sow roots but rutabagas and turnips may be sown as late as midsummer. Root crops are usually planted in drills from 24 to 30 inches apart, with 6 or 8 seeds per foot of drill. Then later the plants are thinned so that they stand from 6 to 8 inches apart in the drill. The cultivation may be done by wheel hand hoes when the plants are small and later by horse cultivators. Some hand weeding, thinning and hoeing must always be done. Clean and thorough cultivation is necessary.

Figure 40.—Cutting vegetable roots for feed.

Harvesting.—Root crops must be harvested before there is danger of freezing weather. Ordinary white frost does not injure root crops. Root diggers may be used, or a furrow plowed with a common plow beside a row of roots greatly reduces the labor of pulling. Roots must be topped usually by hand and stored where they will not freeze but where the temperature is comparatively low.

Questions:
1. Tell if you can why the crops mentioned in this lesson are called root crops.
2. Under what conditions is it advisable to grow roots for stock feed?

3. Describe a good method of growing root crops.

Arithmetic:

1. If a farmer has ten cows and wants to provide each cow 20 lbs. of roots per day for 200 days, how many tons of roots must he produce?

2. If a farmer keeps 40 mature sheep and wishes to feed each 2 lbs. of roots per day for 200 days, how many tons will he need?

3. If roots are planted in rows 24 inches apart and a 2-lb. root produced every 8 inches in the row, how many tons would be produced per acre?

Figure 41.—Sugar beets and a sugar beet factory.

CHAPTER VI

HAY AND PASTURE CROPS

HAY CROP

Importance.—According to the Bureau of Statistics, the annual production of hay in the United States from 1902 to 1911 averaged 60,737,000 tons, grown on 42,557,000 acres, and valued at $624,664,000, an average of $10.28 per ton. The total amount was slightly in excess of the average annual value of cotton or wheat for the same years. The average acre value was $14.72. In addition to its money value, the hay crop bears such an important relation to soil productivity and to live stock enterprises that at least some of the principles of its growth and value should be thoroughly understood by every tiller of the soil.

Advantages.—The value and advantage of the hay crop is often underestimated. Probably you have noticed that it is not necessary to plow and prepare the land for a hay crop as is done for other crops. The grass seed is sown with some preceding grain crop. So seeding of hay costs nothing but for the seed.

If you go out into a good meadow of tame hay at haying time, you will find very few, if any, weeds; and if there are weeds, they will be cut with the hay crop before they produce seed, as hay is usually cut before most of the common weeds produce seed. For this reason the hay crop helps to clean the land of weeds.

Another advantage is that a hay crop makes the soil better for succeeding crops, which is not true of grain or corn crops. If there is clover in the hay, it adds nitrogen to the soil; and any hay crop increases the amount of vegetable matter in the soil, because it has a heavier root system than have any of the other classes of crops. You can prove this by trying to pull a handful of grass in the meadow and a handful of grain in the grain field.

Cost.—There is no other kind of winter feed grown on the farm that can be produced so cheaply in proportion to its feeding value as hay. Some farmers hesitate to devote much of their land to growing hay, because it seems to bring in less money per acre than other crops appear to produce. The fact that it costs much less per acre to raise hay than to raise corn or any of the grain crops, is often overlooked. The following table shows the comparative cost of growing an oat crop and a hay crop. The figures are averages of accurate records on eight farms at Northfield, Minn., for five years.

Cost per Acre, Exclusive of Rent, of Producing Hay and Oats at Northfield, Minn. Average for Five Years.*

Hay—Timothy and Clover.
First Crop.

Seed	$ .410
Mowing	.432
Raking	.165
Tedding	.142
Cocking and Spreading	.177
Hauling in	1.097
Machinery Cost	1.171
General Expense	.623
	4.217

Second Crop.

Mowing	.405
Raking	.156
Cocking and Spreading	.218
Hauling in	.738
General Expense	2.004
Total Cost	6.221

Oats

Seed	1.319
Cleaning Seed	.037
Plowing	1.618
Disking	.397
Dragging	.340
Seeding	.313
Cutting	.473
Twine	.189
Shocking	.193
Stacking	.772
Threshing (labor)	.389
Threshing (cash)	.720
Machinery cost	1.006
General Expense	.804
Total Cost	$8.570

*Minnesota Experiment Station Bulletin No. 145.

HAY AND PASTURE CROPS

The previous table shows that there is approximately twice as much labor and expense in growing an oat crop as in growing a hay crop, even when two cuttings of hay are made. So it is not necessary to get so much from the hay crop as from the grain crop to make as large profits.

Rotation of Crops.—If you can find in your neighborhood a timothy and clover meadow seeded last year, and one on similar land that has been seeded down—that is, raising hay for several years—you will see that the new

Figure 42.—Breaking sod with a traction plow.

meadow, if a good stand has been secured, will give a larger yield than the old. Likewise a meadow or pasture plowed up will usually raise a larger crop of corn or grain than will a field that has not been in grass for several years. These facts show that both the meadow and the grain and corn crops will yield more if new meadows are seeded each year and old ones plowed for other crops. This means rotation of crops, and illustrates an advantage of the practice.

Questions:
1. Name some of the advantages of the hay crop.
2. How does the cost of raising oats and hay compare?
3. How does the yield of hay from an old meadow compare with the yield from a newly seeded meadow.
4. What can you say of the rotation of crops?

Arithmetic:

1. One acre of clover and timothy will produce 2 tons of hay. How much does it cost per ton, if it cost $6.22 per acre to raise it? How much does it cost per ton, if $4.00 per acre is added for land rent?

2. If hay yields but one and one half tons per acre, how much does it cost per ton, if it costs $6.22 per acre to raise it? How much does it cost per ton if $4.00 per acre is added for land rent?

3. If clover hay is worth $8.00 per ton compared with bran at $20.00 per ton, how much is bran worth when clover hay is worth simply the cost of growing it?

CLOVER

As clover is one of the most valuable field crops, it is important that every one know something of its habits and of the conditions favorable to its growth.

Varieties.—There are several varieties of clover, but only four of the varieties are important in the Central West. These are Mammoth, Medium Red, Alsike and White.

Mammoth Clover is very much like medium red. In fact, it is very hard to distinguish one from the other. The Mammoth is much coarser than the medium red, and on that account does not make so good a quality of hay. Its chief value is as a green crop to plow under, though it is often used for hay, pasture or seed.

Medium Red is easily distinguished from alsike and white clover, as it is larger and each leaflet is marked by a V shaped, lighter colored streak near its center. The red blossoms aid, also, in distinguishing this variety of clover. It will be noticed that nearly or quite all the stems of this clover are covered with fine hair. These hairs are objectionable, as they have a tendency to gather dust, thus making dusty hay, unless very carefully cured. It is usually a biennial, that is, as a rule, it lives but two years. It is usually sown with some grain crop, called a nurse crop. The clover plants are very small during the early part of summer, and do not grow much until the grain crop is cut. During the fall the clover grows very rapidly; but it does not produce a crop until the next year, the second year of its growth, when it produces two crops—two hay crops or a hay and a seed crop. The second crop is the one usually saved for seed. After the two crops are cut, the plants usually die, as they have lived their life. An occasional plant may live over, and considerable clover may appear in the field the third year; but this is largely due to seeds formed the first or second year of the clover's growth, or to clover seeds starting that failed to grow the first year. This variety of clover is especially adapted to rotation pastures and meadows. Considerable trouble is experienced in curing this variety of hay, as the thick stems contain so much juice that in trying to dry them the leaves, which are very thin, are liable to become too dry and shatter off when the hay is handled.

Figure 43.—Leaves and stems of clover. 1. Medium red. (Note markings on leaflets, also hairs on stems.) 2. White. (Note smooth stem and small leaflets.) 3 Alsike. (Note smooth stem and smooth leaflets with prominent midrib.)

Alsike Clover is smaller than medium red clover, has smaller, more oblong leaves without white markings, and there are no hairs on its stems. The blossoms are smaller and lighter colored, nearly white at first and later pink. The seed is much smaller and darker colored than the seed of medium red. Alsike clover makes a better quality of hay than the red clover, because it is free of hairs and finer; but, as a rule, it does not yield so much on upland. It is a perennial, that is, it lives for several years unless some unfavorable conditions kill it. On this account it is more valuable for permanent pasture or meadow than the medium red. It is quite well adapted for low wet places, as it will stand more water than the red clover.

White Clover is a very small, low growing plant with a tendency to trail along on the ground. It has small, nearly round, smooth leaves and smooth stems. The seed is a little smaller than alsike clover seed and is yellowish in color. It is a perennial. The stems creep along on the ground and take root at the joints, thus starting new plants. It spreads in this way as well as by the seeds. The blossoms

94 ELEMENTS OF FARM PRACTICE

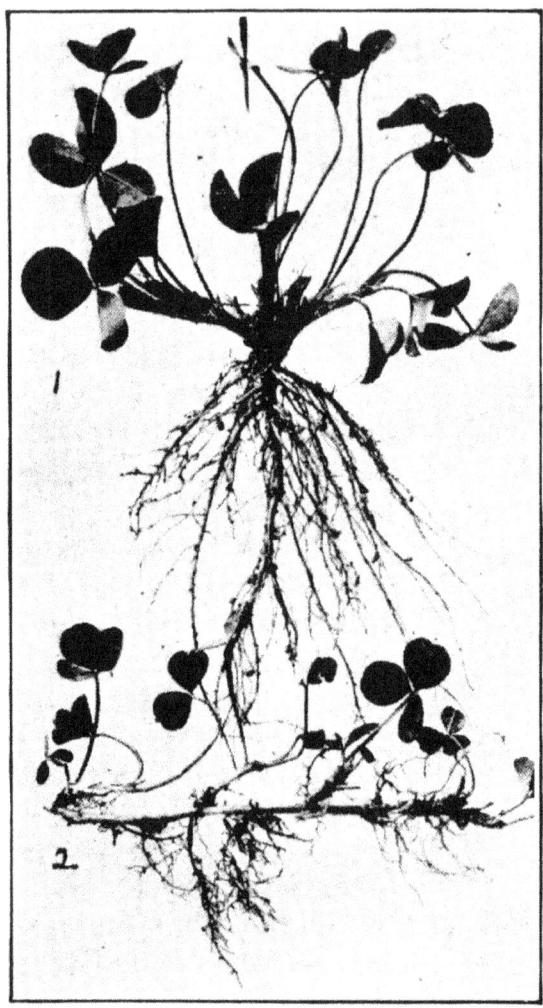

Figure 44.—Root systems of alsike and white clover plants. 1. Alsike. (Note small taproot and comparatively large laterals.) 2. White. (Note creeping stem rooted at different places, and fine fibrous roots.)

are white, and they can usually be seen at any time during the summer from June 1st until it freezes up in the fall. The white clover is of very little value as a hay crop, as it grows too short. It is common on lawns and in old pastures and is a valuable plant for such places.

Getting a Catch of Clover. — Difficulty is sometimes experienced in getting a good catch of clover. As clover grows slowly the first year it is sown, the grain crop with which it grows is liable to crowd it pretty hard; and when the grain is cut the hot sun is pretty hard on the small plants, especially if the weather is dry. Clover seed should be sown only on a fine, mellow rich, well-prepared seed bed. Land well manured, planted to corn and well cultivated, then well disked the following spring, and sown to grain, is in fine condition for clover seed.

If clover seed is sown on poor land the chances of getting a catch are greatly increased if a light dressing of manure can be applied soon after the seed is sown.

Questions:
1. Name the four varieties of clover common in the Middle West.
2. Which two are very much alike?
3. Describe the leaf, stem and blossom of each of the last three.
4. What are annual, biennial and perennial plants?

HAY AND PASTURE CROPS

Arithmetic:

1. If an acre of clover yields 3,500 lbs. of hay at the first crop, and a bushel of seed at the second crop, what is the value of the entire crop, hay $10 per ton and clover seed $8 per bushel?
2. If an acre of clover yields 200 lbs. of seed, what is its value at $8 per bushel? (60 lbs. per bu.)
3. If clover seed is worth $8 per bushel, what is its value per hundred weight?

CLOVER ROOTS AND BACTERIA

Medium Red Clover Root.—An examination of the roots of medium red, alsike and white clover will show considerable difference in them. The medium red clover has a large taproot (a root running straight down in the soil). This root is much larger and longer than the root of either of the other clovers. If a plant is dug carefully from a well drained soil, this taproot will be found to extend down two, and often more, feet; which shows that this clover gets part of its food from the subsoil. It has also many lateral roots running out from the taproot. In fact, if roots are carefully taken up, it will be seen that there is nearly as much plant below as above ground. On this account medium red clover is one of the very best crops to grow to add vegetable matter to the soil.

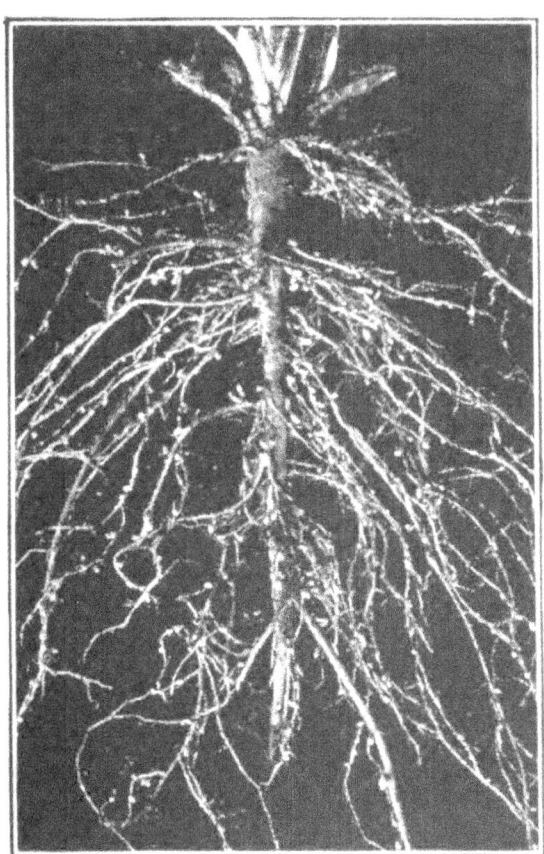

Figure 45.—Root system of medium red clover. (Note large taproot, lateral roots and nodules caused by nitrogen-gathering bacteria.)

Alsike clover roots are considerably smaller than those of medium red clover. In many plants the taproot is not plainly seen or is not much larger than the fibrous roots.

The roots of the alsike clover extend into the soil to considerable depth, however, thus enabling the plants to draw on the subsoil to some extent for plant food. This clover also adds a large amount of vegetable matter to the soil by its roots. Both alsike and medium red are very beneficial to heavy soils, by opening them and letting in air when the roots decay. They are beneficial to sandy soils by adding large amounts of vegetable matter, thus making the soils capable of holding more moisture.

White clover roots are very small and fibrous. No taproots are found, and the fibrous roots do not go nearly so deep as the roots of the other clovers. The plants grow so thickly, owing to their habit of spreading, that they thoroughly cover the ground and keep the surface soil well supplied with vegetable matter.

Clover Adds Nitrogen to the Soil.—Clover possesses, beside its heavy root system, another feature which makes it a valuable crop to improve the soil. If a clover plant is carefully dug from the soil, small bunches or nodules about the size of an ordinary pin head will be seen on the roots. See Figure 45. These are caused by bacteria. Bacteria are a very low form of plant life. They are unable to live from the soil as higher plants do, but must depend upon plant or animal substances to supply them with organic matter. Disease germs, the germs that cause milk to sour, the germs that cause decomposition or rotting, etc., are also bacteria. Some bacteria live on dead matter, others on live matter. The latter are called parasites. The bacteria causing the nodules on clover roots are in a sense parasites, but in this case they are beneficial; they do something for the clover plant that it is unable to do for itself. All plants require a large amount of nitrogen for food. A very large proportion of the air is free nitrogen. Our common field crops are unable to make use of this nitrogen; but clover, alfalfa, peas, beans and other plants belonging to the family called legumes have the habit, which no other class of plants has, of forming a sort of partnership relation with these bacteria and through them are enabled to draw upon the nitrogen of the air. These nitrogen-gathering bacteria have the power to absorb the nitrogen from the air and to pass it on to the plants on which they are growing. In this way a soil lacking in nitrogen may be made richer in this element by growing a legume crop. This is true even though the crop be removed from the field, as the roots and stubble left are rich in nitrogen. Nitrogen, when bought in commercial fertilizers, costs about 18c. per pound. A farmer, by growing clover or some other legume crop, can add enough nitrogen to the soil to grow several crops of corn or grain, besides his legume crop.

Questions:
1. Which variety of clover, medium red or alsike, has the heavier root system?
2. In what ways is a clover crop beneficial to the soil?
3. What enables clover and other plants belonging to the same family to make use of the free nitrogen in the air?

Arithmetic:
1. What is the value of 150 lbs. of nitrogen at 18c. per pound? (Note: An acre of clover may add 150 lbs. of nitrogen to the soil.)
2. How many crops of wheat, each crop removing 25 lbs. of nitrogen per acre, would use the nitrogen added by a crop of clover?
3. If an acre of clover yields 3,500 lbs. of hay the first cutting, and 2,500 lbs. the second, what is the value of the hay at $10 per ton?

CURING HAY

The Weather.—The quality of hay and its value as food depends very largely on the way it is cured. Since hay on a great many farms forms a large part of the winter food for stock, it is important that it be cured in the best possible way. The weather has a great deal to do with the curing of hay, and some seasons it is practically impossible to get hay well cured. But there are certain principles involved that, if followed, will usually result in a better quality of hay than is secured by methods commonly followed. The suggestions given below apply to clover, but, if followed, will give good results with any heavy crop of hay.

Time to Cut.—While hay that is cut when quite ripe yields more per acre and is easier to cure than earlier cut hay, it is much less digestible, less palatable, and contains a smaller proportion of protein, which is the most valuable and costly element in hay. Experiments show that the greatest amount of digestible food is secured when hay is cut at about the time it is in full bloom. With clover this is usually from about June 15th to July 1st. Never cut hay while the dew is on it, for time is lost in the drying. The dew will dry off more quickly while the hay is standing.

Curing.—If clover is cut in the forenoon of a bright day, it should be turned over, with either a rake or a tedder, before any of the top leaves become dry. The object sought in curing clover hay should be to keep the leaves green as long as possible, as they help to draw the moisture out of the large stems, which are the difficult part to cure.

Figure 46.—Cock covers in use at the Minnesota Experiment Station in curing alfalfa hay.

If it gets dry enough the first day, so that a good job of raking can be done, rake it before night; if not, ted it, if possible, so as to get the green hay from the bottom on top to take the dew, as dew will blacken partly cured clover. If it looks like rain, cock the hay as soon as possible; if not, leave it in the windrow. The next morning as soon as the top part of the hay is nearly dry, rake, or, if raked, turn the windrow over either by hand with a fork or with the team and rake. Aim to keep the hay loose in the windrow, so that the air can pass through it freely. The leaves are largely protected from the hot sun in this way, and can perform their function of drawing water from the stems, and are not shattered off and lost. It is usually wise to cock the hay the second day, if it is too green to store, and leave in the cocks a day or two; then open up cocks for an hour or so to the sun and wind; then put under shelter, either in barn or stack.

Damaged by Hot Sun.—The old adage, "Make hay while the sun shines," is good advice, but may be overdone. Hay, especially clover, that has been exposed to the sun for very long is very materially injured, as the thin leaves

Figure 47.—The hay sling in operation in unloading hay. Two or three sling loads will take off a large wagon-load of hay and leave very little scatterings.

are dried up and lost and little is left but stems. The heat of the sun is very essential to evaporate the moisture from the leaves and stems. This may be accomplished, however, by curing the hay in the cock or windrow and without allowing the sun to shine directly on the leaves.

Cock covers may often be used with profit. It seems expensive when one first considers them, but if one considers that bran is worth $20.00 or more per ton and that good clover hay is worth very nearly half as much as bran for feed, one must acknowledge that the difference in value between good and poor hay is often more than the cost of using covers.

Cock covers are pieces of canvas or sheeting about four feet square, with weights sewed in each corner, so that when one is spread over a hay cock, the wind will not blow it off. Such a cover helps to protect hay, while curing, from both the sun and the rain.

We would advise our readers to test curing clover largely in the shade, as suggested above, with at least one cock of hay, then compare with other clover cut at the same time but exposed to the sun and dew for two or three days.

If clover is cured until thoroughly dry, and then cocked, it will shed very little water, as the stiff stems stick out in every direction and the water follows them down through the cock. But, if hay is cocked when only partly dry, the stems are limber and wilted and so hang down over the sides of the cock, and tend to shed water.

Questions:
1. In what ways may the value of hay be reduced in curing?
2. At what stage of growth should a hay crop be cut? Why?
3. What is gained by protecting clover hay from the sun when curing it?
4. Do you think it will pay to use cock covers in curing hay? Why?

Arithmetic:
1. If a cock cover large enough to cover 80 lbs. of hay costs 20c. how much will it cost for enough to cover 1 ton of hay?
2. If cock covers can be used five times each year and will last 5 years, how many times can each be used during its lifetime?
3. If enough cock covers to cover a ton of hay cost $5.00 and can be used 25 times, how much does it cost per ton for cock covers?

ALFALFA

Alfalfa has been grown for forage for many hundreds of years. It is not a new crop, though in many sections of the country it is just being introduced. Alfalfa and corn are the two crops that have been most talked about during the past ten years, and on this account many persons regard alfalfa as a new crop.

Advantages.—The chief advantages of alfalfa over other hay and pasture crops are its richness in food value, the rapidity of growth,—it produces from two to four crops per year,—and its perennial habits. Common clover usually grows but two years, and furnishes a crop but one year. When alfalfa is once established in a field, it remains and continues to produce good crops for many years. Alfalfa is also a valuable crop to enrich the soil on which it grows. Like clover, it has on its roots bacteria, which gather the free nitrogen from the air and eventually add it to the soil.

It has an exceedingly heavy root system. The roots grow into the soil to a greater depth than any other common cultivated crop. When an alfalfa field is broken up, the soil is rich in nitrogen, and well supplied with vegetable matter. As alfalfa grows rapidly and is cut two or more times a year, it is a very good crop to rid the soil of weeds.

Feeding Value.—Alfalfa is the richest hay crop produced, and under favorable conditions produces more food value per acre than any other hay crop. The following table shows the composition of alfalfa hay as compared with other common kinds of hay.

Composition of Common Kinds of Hay Nutrients in One Pound.

	Protein Pounds	Carbohydrates Pounds	Fat Pounds
Alfalfa	.11	.40	.012
Alsike clover	.84	.42	.015
Medium red clover	.071	.38	.018
Prairie hay	.030	.42	.014
Timothy hay	.028	.43	.014
Slough hay	.026	.42	.011
Fodder corn	.025	.35	.012

Besides being richer than other kinds of hay, alfalfa is relished by all classes of live stock, including horses, cattle, sheep and swine. When well cured and intelligently fed, it is a most valuable feed for all farm animals.

Pasturing.—Alfalfa pasture is among the best pastures. All kinds of stock eat it readily. In fact, it is relished so much that great care is necessary in turning stock into an alfalfa field to prevent them from bloating. Stock should never be turned on alfalfa when they are hungry or when the alfalfa is wet. There is no better hog pasture known than alfalfa, the main difficulty being that the hogs are likely to root up the alfalfa and thus thin out the stand. If given large pastures and plenty of salt and ashes, hogs are much less likely to injure the alfalfa by rooting.

Seed.—Alfalfa produces seed readily, especially in rather dry sections. One condition that has prevented a rapid increase in alfalfa in the Northwest has been the difficulty of getting seed that is entirely hardy. Imported seed, or seed grown farther south in the United States, is not sufficiently hardy to live over winters where the climate is severe. Quite a number of men have been trying for years to get or produce hardy strains of alfalfa for the Northwest, and it is now possible to get seed that is comparatively hardy. It is advisable when possible to get alfalfa seed from a field that has grown under conditions similar to your, own and has withstood the winters. Such seed is reasonably sure to be hardy. It is never wise to sow alfalfa seed that you have purchased until you have had it tested for germination and for purity. You want to be sure the seed will grow, and that it contains no bad weed seeds. Your experiment station will usually make these tests free of charge.

Soil for Alfalfa.—With hardy seed, alfalfa may be grown successfully on nearly any soil that will produce good corn. It requires a well drained soil and soil that is not sour. The richer the soil is, the better the alfalfa will grow; but, when once started, it will make fairly good growth on even poor soils. If soil is sour, the application of limestone will correct it. It is advisable, if possible, to plow land for alfalfa in the fall, then top-dress it with manure, either in the fall or spring, then in the spring thoroughly disk and harrow the land before the alfalfa seed is sown.

Sowing.—Alfalfa may be seeded in the spring with a nurse crop or in midsummer alone. The safer plan is to disk and harrow the field every few days from spring until midsummer. These operations will destroy most of the weeds, make a very fine mellow seed bed, and leave the soil warm, moist and rich. Seed sown under these conditions will start and grow rapidly, and usually get ahead of the weeds. Weeds are quite troublesome in getting alfalfa started. When sown in midsummer the seed is usually sown broadcast and covered with the harrow. If sown in the spring, it is usually sown with a nurse crop the same as timothy or clover.

Inoculation.—Alfalfa, like clover, has the power of adding nitrogen to the soil. It can do this only when it has alfalfa bacteria growing on its roots. The necessary bacteria are not always present in the soil. This is likely to be true in soils that have never grown alfalfa before. Farmers have found this out and are attempting to provide the bacteria artificially. This may be done in several ways. The most common way is to apply a few hundred pounds of soil from an old alfalfa field to the newly seeded field. This must be done without exposing the soil handled very much to the sunlight, as the sun will quickly destroy the bacteria. The United States Department of Agriculture will furnish the bacteria free in a small bottle, together with instructions for using. This is a very easy way of making sure that the proper bacteria are present.

Figure 48.—Cutting a field of alfalfa.

Cutting.—Alfalfa may be cut two or more times each year according to soil, climate and degree of development desired. It should be cut when the shoots representing the new growth have started. This stage is easily determined by looking at the plants near the ground. If at any time the crop begins to turn yellow, it is advisable to cut it, as it will make no further growth of value until cut.

Curing the Hay.—Alfalfa hay, like clover hay, is quite hard to cure in the parts of the country where there is likely to be rain. Most of the alfalfa is growing in the semi-arid or irrigated sections where there is little rain to bother. In the states where clover is grown, alfalfa hay is handled in about the same way as clover hay; that is, it is cut when the second growth starts which sometimes is before it blooms, then raked as soon as it can be raked, and cured largely in the windrow or small cock. This is to save the leaves from drying up and shattering off, as the leaves are the

Figure 49.—Curing alfalfa hay under cock covers.

richest and most valuable part of the hay. As alfalfa hay is worth about $15 per ton for feed, it usually pays to use cock covers in curing it to protect it from the rain and dew.

Questions:
1. In what way does alfalfa differ from medium red clover?
2. What classes of stock will eat alfalfa hay?
3. How would you prepare a field for alfalfa?
4. What do you understand by inoculating soil for alfalfa?

Arithmetic:
1. Fifteen pounds of alfalfa seed are required to seed an acre; how much will the seed cost at 20c. per pound?
2. If an acre of alfalfa produces 3 crops in one year, 1st crop 3,500 lbs., 2nd crop 3,000 lbs., and 3rd crop 2,500 lbs., how many tons would be produced.
3. If a ton of alfalfa hay is worth $10, how much would the hay produced on the acre mentioned in Example No. 2 be worth?

OTHER COMMON HAY AND PASTURE CROPS

Timothy is the most common hay crop grown. It is grown in every state in the Union. It is a perennial plant, and will continue to produce hay or pasture year after year without reseeding, unless the land is plowed up. It may be grown alone or mixed with other grasses. A very common grass mixture for seeding meadows or pastures is: 6 pounds medium red clover and 8 pounds of timothy seed per acre. When timothy is sown alone, about 10 pounds of seed are required. The seed is usually sown with a nurse crop such as wheat, oats or barley. The grain crop protects the young plants from the hot sun and dry winds, and also prevents the growth of weeds. The grain crop also produces on the land a valuable crop during the year the grass crop requires to get started. In this way no time is lost, and a crop is secured every year.

Cutting for Hay.—Timothy usually produces but one crop of hay each year, though sometimes a light second cutting is secured. Like clover, it should be cut when in bloom, but it does not bloom until about two weeks after red clover blooms; so, if the two are grown together, one crop must be cut a little too early or the other a little too late. Timothy is cut and cured very much the same as clover; but it is much easier to cure; because it does not grow so heavy or have such large juicy stems. In favorable weather it may be cut in the morning as soon as the dew is off and raked up the same afternoon.

Value for Feed.—Owing to the ease of growing and curing, and its freedom from dust, timothy hay is the favorite for horses. If cut in proper season it is readily eaten by all classes of animals. It is not popular, however, as feed for cattle, chiefly because it is not rich in protein or muscle-forming material.

Timothy Pasture.—Timothy is very generally used for pasture, but it is not so valuable for this purpose as for hay. It is used because seed is cheap, and because meadows that have produced hay for a year or more are very commonly used for pasture. For one or two years timothy furnishes fairly good pasture, but after that the stand gets thinner and rather bunchy. If it is to be used for pasture, it is well to seed with the timothy alsike clover and if one is seeding down a permanent pasture, that is a field to be in pasture five or more years, it is better to add five pounds to ten pounds per acre of Kentucky blue grass seed. The timothy and clover will furnish the greater part of the pasture for the first two or three years, and after that the

Kentucky blue grass will gradually take its place. After blue grass gets a good start, it furnishes much better pasture than timothy. Permanent timothy or blue grass pastures are greatly improved by harrowing and top-dressing, (spreading on a light coating of manure), every two or three years.

Brome grass is another valuable hay and pasture crop. It is unpopular in many places because many farms have been infested with quack grass from seeding brome grass. The seed of quack grass and brome grass are so much alike

Figure 50.—Haying time.

that only an expert can distinguish between them. Many of the states now have seed laws that require that all seeds for seeding purposes must be labeled, and if a sample of seed contains quack grass it must be so stated on the label. Most of the states have seed laboratories in connection with their experiment stations in which samples of seed will be tested free. These laws will make it safe to sow brome grass. Brome grass is very well adapted to conditions where there is likely to be a shortage of rainfall. It is especially valuable in such places for pasture. It will probably never be a very important crop where timothy, clover and alfalfa are easily grown. Its feeding value is about the same as timothy, and it is cured and handled in about the same manner. From ten to fifteen pounds of seed are sown per acre, usually with a nurse crop.

HAY AND PASTURE CROPS

Other Varieties of Grass.—Kentucky blue grass, redtop, orchard grass, Johnson grass, rye grass, etc., are other important and valuable grass crops for special conditions. Space here is too limited to discuss them. Bulletins from your state experiment station, or from the United States Department of Agriculture, giving full information about these grasses, may be secured by writing for them.

Questions:
1. Why is timothy such a popular hay crop?
2. Tell what you can about brome grass.
3. Why are the grass crops mentioned in this lesson of less value than timothy?

Arithmetic:
1. If one sows 6 lbs. of red clover seed worth 15c. per lb. and 8 lbs. of timothy seed worth 6c. per lb., how much does the seed cost per acre?
2. If an acre of timothy produces 1½ tons of hay, how long will it last a horse fed at the rate of 15 lbs. per day?
3. It takes 15 lbs. of brome grass seed to seed an acre. How much does it cost at 18c. per lb?

CHAPTER VII

MISCELLANEOUS CROPS

FORAGE CROPS

Millet is commonly grown as a catch crop where some other crop has failed, or where it is necessary to sow a crop late. It is usually cut for hay, but some of the varieties known as the broom corn millets are grown for seed. The seed is used as feed for live stock. Millet has about the same effect on the soil as a grain crop. It is very good as a cleaning crop, as it grows quickly and covers the ground very thoroughly. It will grow on most any kind of soil. It may be sown any time during the early summer, and will be ready to cut for hay eight or ten weeks after seeding. Two to three pecks of seed per acre should be sown.

Rape is a forage plant that appears, when growing, very much like rutabagas. It does not, however, produce a heavy edible root like the rutabaga. It is used chiefly as a pasture plant for hogs and sheep. Cattle will eat it, but

Figure 51.—A field of rape ready for sheep or hogs.

it is not regarded an important pasture crop for cattle. It is often sown with grain crops in the spring at the rate of from one to three pounds of seed per acre. After the grain crop has been harvested the rape grows up in the stubble and furnishes pasture for stock during the fall. For small fields to be pastured by sheep or hogs, it is commonly sown alone, at the rate of three or four pounds per acre. It may be seeded at almost any season of the year until fall. Six weeks after seeding it is ready to be pastured off, if fair growing conditions have prevailed. It is one of the most popular annual crops grown for hog and sheep pasture, because the seed is comparatively cheap.

Field peas are grown to some extent for seed and also for hay and pasture, especially in the North. Its chief value is for pasture for hogs and sheep, either when green or when the crop is mature. The chief objection to the crop is the amount of seed required per acre and its cost. Peas must be sown early in the spring. They may be sown alone at the rate of three to four bushels of seed per acre, but are more commonly sown with oats at the rate of two bushels of peas and one bushel of oats. The crop may be pastured off green, cut for hay, or allowed to ripen and then fed off by hogs or sheep.

The soy bean is an annual legume crop quite commonly grown in southern parts of the United States, but at present is not of great importance in the North. The common field pea takes its place in the North, because the soy bean is very tender to frost. This crop is grown for seed, for hay and for pasture. It is sown both in drills and broadcast. When sown in drills about one half bushel of seed is required per acre, and when sown broadcast about one bushel of seed is used. It is not sown until danger of frost is past.

The cowpea is another annual legume that is becoming of great importance in the South as a green manure hay and pasture crop. It may be sown late in the summer after a corn or cotton crop is out of the way, or between the rows of corn or cotton, then cut for hay, pastured or plowed under late in the fall. It, also, is very tender to frost, and is not profitable as far north as is the soy bean.

It may be sown in drills and cultivated, or broadcast, as desired. When sown in drills from two to three pecks of seed are used; when sown broadcast from four to six pecks.

Vetch is an annual legume plant, very fine and trailing in character. It is grown for hay, or as a green manure crop. There are two kinds, spring and winter. The winter, or hairy vetch, is by far the more valuable. It may be sown in the fall or spring, as desired. It is commonly sown in the fall with rye, and the whole crop cut for hay early the following summer or plowed under. It is sown at the rate of from four to six pecks per acre. None of the annual legume crops are important in general farming where the clovers and alfalfa can be grown successfully.

Questions:
1. Tell what you can about the uses and culture of millet.
2. Tell what you can about the uses and culture of rape.
3. Tell what you can about the uses and culture of the annual legume crops discussed in this lesson.

Arithmetic:
1. If rape seed costs 8c. per pound, and 3½ lbs. are required per acre, how much does rape seed cost per acre?
2. If 3 bus. of peas are required to seed an acre, how much will they cost at $1.50 per bushel?
3. What is the value of an acre of rape pasture, if it will produce as much pork as 1,000 lbs. of shorts, when shorts are worth $25 per ton?

RICE AND SUGAR CANE

Rice is one of the oldest cultivated plants, and forms the staple article of diet for millions of people in India, China and Japan. The world's rice crop exceeds the world's wheat crop or corn crop. Nearly all the rice grown in the United States is raised in three states, Louisiana, Texas, and Arkansas. The fertile river valleys and plains of these states and their warm climate make rice-growing profitable.

The land is prepared in much the same manner as for other grains,—plowed in the spring and disked and harrowed. The seed is generally sown with a grain drill at the rate of one to two bushels per acre, either the latter part of April or the first of May. When the crop is about eight inches high, it is flooded with water to a depth of from three to six inches, which depth is maintained until the crop begins to ripen. The water is then drawn off to allow the ground to dry for harvesting. The crop is cut with a binder, stacked, and threshed as other grains. In milling rice, it is not ground, but is hulled and the kernels polished. Rice is more nutritious if it is not polished, as this process removes the portion which contains the fat.

Sugar Cane.—Sugar cane is a plant which greatly resembles corn, only no ears are produced. It is grown for the juice contained in the stalk. Sorghum, raised in small quantities in the northern states to produce molasses for family use, is an annual; but the sugar cane of the southern states is a perennial. The former is raised from seed which forms on top of the stalk. The latter is propagated from sections of the stalk.

Soil and Climate.—Ordinary good soil is suitable for sugar cane. In regions of slight rainfall, irrigation is necessary. The crop requires a long, hot season.

Planting and Cultivation.—Sugar cane is generally planted by laying the entire stalk or a portion of it in furrows from four to six feet apart. The new plants grow from the buds at the bottom of the leaves. Frequent cultivation is necessary, also hand hoeing.

Harvesting and Manufacturing.—The canes are first stripped of their leaves, and then cut off close to the ground with a knife. As they soon begin to lose their juice, it is important that they reach the mill as quickly as possible. At the mill the stalks are first shredded and then passed between heavy rollers. The crushed stalks are used in the mill furnace. The juice is first purified and filtered, and then boiled to sugar crystals.

Questions:
1. How does rice compare in importance as a world crop with corn and wheat?
2. State briefly the methods used in growing rice.
3. Tell what you can of the use and methods used in growing sugar cane.

Arithmetic:
1. If Louisiana produces 380,000 acres of rice yielding 1,000 lbs. of cleaned rice per acre, how many pounds of rice are produced in the state?
2. If the world's production of cleaned rice is 175,000,000,000 lbs., and the production in the United States is 550,000,000 lbs., what per cent of the world's rice crop is produced in the United States?

3. The world's production of cane sugar amounts to about 8,000,000 long tons, of which continental United States produced about 325,000 tons. What per cent of the world's crop was produced in the United States?

FIBER CROPS

Fiber is a slender, threadlike substance used in making numerous things, such as cloth, silk, rope, thread, twine, paper, etc. It is obtained from two sources, animal and vegetable. The most important animal fiber is wool from the sheep and the silk spun by the silk worm. A large part of the vegetable fiber is taken from the three plants, cotton, flax and hemp.

Cotton.—Cotton is a plant raised extensively in the southern states. The large, white blossom of this plant turns pink the second day, and later develops into a body about the size and shape of an egg. When ripe, this breaks open, exposing the seeds which are covered with fiber or lint from one to two inches long. It is this lint when removed from the seed that is used in making thread, cloth, etc.

Soil.—The largest crops of cotton are raised on the rich, loamy soils of the southern Mississippi Valley and the clay loams of Texas. But it may be raised successfully on light sandy soils, if they are fertile and moist. (Nearly all the cotton production in the United States is limited to the southeastern states.)

Planting and Cultivation.—The seed is usually dropped evenly in furrows. The furrows are generally from two and a half to five feet apart. When plants are about two inches high the spaces between rows are plowed, and the plants in the rows thinned out with a hoe until they are usually from one to two feet apart. The crop is later cultivated from three to five times, and hoed once or twice.

Picking and Ginning.—The cotton picking is done by hand, and commences as soon as a considerable number of pods or bolls open. Usually three pickings are necessary.

In the ginning the lint is removed from the seed and packed into bales ready for the mills.

Flax.—The flax plant is an annual which grows from twelve to twenty inches high. It has a single, upright stem and a light blue blossom. It is from the stem that the fiber is taken. Flax is raised also for the seed or grain from which oil is made. After the oil has been removed, the remaining part of the grain is used to feed stock. The oil is called linseed oil and is used extensively in mixing paints.

Soil.—Flax grows best on a comparatively light soil. Sandy loams are better than clay. It grows better than any other crop on tough sods. For that reason it is often the first crop on newly cultivated land.

Preparation and Planting.—On old land, deep plowing and thorough packing of seed bed are necessary. Sod land is usually plowed in the fall or early spring. The plow is run just deep enough to turn the sod over. It may then be disked and packed by a roller. The seed is usually sown with a grain drill from one to two inches deep. If flax is raised for fiber, it is seeded thickly, as this method produces longer stems with fewer branches and less seed. About two bushels to an acre is the average rate of seeding for fiber. If flax is raised for seed, it is thinly sown so that the plants may branch freely, thus producing more seed. From two to three pecks to an acre is the average rate to produce seed. It may be sown as soon as danger of frost is over, usually about the middle of May, and is harvested about the first of September.

Harvesting.—Seed flax is cut with a grain binder, or with a reaper, and threshed much the same as wheat. Fiber flax is usually pulled by hand, tied into small bundles, and put into shocks to cure. Two or three weeks later the seeds are rubbed out, also by hand. The straw is then spread out thinly on the ground and left to weather for three or four weeks. This process is known as retting. The straw is then pounded or bent to separate the fiber. The fiber is then ready for the manufacturer.

Hemp.—Hemp is an annual plant, and grows from eight to twelve feet high.

Soil.—Rich land well fertilized gives the best results.

Culture.—About five pecks of seed to the acre are sown either broadcast or with a grain drill. It usually grows rapidly, and is ready to harvest when the seeds ripen, which is usually about three and a half months from planting time. It is cut with a mower or a corn knife and allowed to lie on the ground to ret. Hemp fiber is mostly used for the manufacture of carpet warp and rope.

Questions:
1. What do you understand by the term fiber crops?
2. Tell what you can about the uses and culture of cotton.
3. Tell what you can about the uses and culture of flax fiber.

Arithmetic:
1. If Texas produces 25% of the 12,000,000 bales of cotton produced in the United States, how much is produced in Texas?
2. If 5/8 of the 12,000,000 bales of cotton produced in the United States is exported, how many bales are exported?
3. If a farmer produces 50 acres of cotton yielding 400 lbs. per acre, how many 500-pound bales will he produce?

CHAPTER VIII

COMMON WEEDS AND THEIR ERADICATION

WEEDS

A Weed is any plant out of place. For example, rye growing in a wheat field or any grain plant growing in a corn field is as much a weed as is pigeon grass. But we commonly think of weeds as undesirable plants that are found in our fields, meadows and pastures, such as mustard, thistle, etc.

Weeds are harmful in many ways. As we have learned in the previous lesson, they use moisture and plant food that are needed by the useful crops. They shade or crowd out other plants. They greatly increase the cost of growing crops. They increase the cost of harvesting, by requiring more twine and by making more bulk to handle. They decrease the quality of grain and increase materially the cost of marketing.

There is no accurate way of estimating the loss caused by weeds, but it is very great. Weeds cost many times as much as all the schools in the country.

One of the great problems of farming is the control of weeds, and no farmer can make much of a success of his business until he learns how to fight weeds effectively.

Weeds get into fields in a great many different ways. Some weed seeds remain in the soil for several years, and still retain sufficient vitality to grow when given favorable opportunity. Weed seeds are carried into fields by water, by wind, by birds, by animals, by machinery, or sometimes in the seed grain or in the grass seed.

There are no very easy ways of controlling weeds, but the first essential of success is to know the common weeds and their habits. Then one may discover the measures necessary for their eradication.

There are not a great many different weeds that are very troublesome, and it is not difficult to become so familiar with most of the common ones as to recognize them when seen, either as seeds or as young or mature plants.

Specimens.—It may prove an interesting and profitable pastime, at the proper season of the year, to gather specimens of all the weeds that you commonly find in your fields, to observe them carefully and try to find some characteristic by which you can identify each kind. You may desire to press, mount and name these specimens.

Figure 52.—Yellow mustard, showing taproot, hairy stem and (1) the seed pod split open; (2) blossom, showing 4 petals in the form of a cross, whence this family of plants is named Cruciferae; (3) Seeds.

Mounting Weeds.—Select two or three plants that represent their class, and dig them up in such a way as to show the root, the leaves, the stem, and, if possible, the blossoms. Lay or hang them in the shade until well wilted but not dry. Then spread out the parts carefully, to show each plainly. Lay the plants between a couple of sheets of blotting paper, if you have them; if not, put the plants between newspapers, and put heavy weights on them. Change the papers often until the plants are dry, to prevent them from molding. When dry, mount them on a piece of white paper by pasting over the stem and branches, and upon the paper at several places, little strips of paper, with mucilage or paste on one side. Plants carefully mounted will be of great value for use in identifying weeds.

Value of Collection.—Handling plants so thoroughly and carefully, as is necessary to gather and mount them, makes one quite familiar with them. You may be sure your teacher would appreciate such a collection of weeds for use in the schoolroom, especially if they are named. If you do not know the name of some weed, and cannot find out in your neighborhood, get as nearly a perfect specimen of the plant as you can (being sure to get the roots, stem, leaves and, if possible, the flowers or head) and send it to your State Experiment Station, and it will be named for you. If you have studied botany or expect to study it, you will find your work with weeds of great value.

It is hoped that every reader will examine carefully the weeds commonly found in his locality, until he can recognize them all at sight.

Questions:
1. What is a weed?
2. What do we commonly think of as weeds?
3. How are weeds harmful?
4. How great is the loss caused by weeds?

Arithmetic:
1. If a field of wheat yielding 18 bus. per acre were injured 10% by weeds, how much would it have yielded had it been free of weeds?
2. If a boy can pull the mustard in an acre of grain in two days, what does the mustard cost the farmer, if the boy's time is worth 60c. per day?
3. If a man spends an hour cleaning enough seed grain for two acres, how much will it cost him per acre, if his time is worth 14c. per hour?

WEED SEEDS COMMON IN GRASS AND CLOVER SEED

Clean Seed Grain.—However careful a farmer may be and has been for several years, some weeds are bound to spring up and grow from roots or from seeds which have lain dormant in the soil for a year, or perhaps longer, until recent plowing or harrowing has placed them where they can grow. But many farmers increase the amount of weeds in their fields, and often introduce new and bad varieties by buying or using seed grain that contains these weed seeds. The weeds in each year's crop may be lessened noticeably by sowing only grain free of weed seeds.

Farmers should be able to recognize the weed seeds found in grain, so that they may not buy and use seed grain that contains seeds of dangerous weeds.

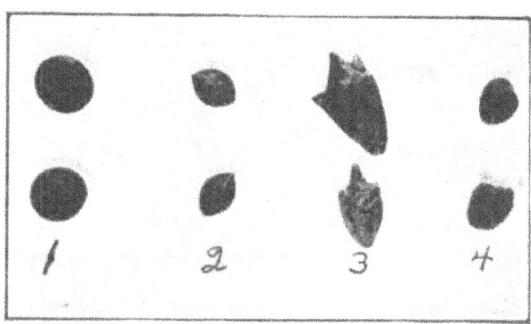

Figure 53.—Seeds of (1) wild pea or vetch; (2) wild buckwheat; (3) Ragweed or Kinghead; (4) corn cockle.

Description.— Below is given a brief description of five kinds of weed seeds most commonly found in grain:

Corn cockle or blue cockle is a rough, black, somewhat triangular seed, about as large and heavy as a kernel of wheat. It is common in seed wheat, as it is hard to separate these seeds from the grain. See Figure 53.

Ragweed or kinghead is a dark brown heavy seed. The seeds vary in size from slightly smaller to considerably larger than a kernel of wheat. They are easily recognized by the crown-like appearance of the tip. The seed is smaller at the base, with several ribs extending lengthwise and terminating in as many points around a central point in the tip, giving it the crown-like appearance mentioned. It is common in grain in the Red River Valley. See Figure 53.

Wild oats may be distinguished from common white oats by the following points: Wild oats are darker in color, are more slender, have a small tuft of hair at the base and have a

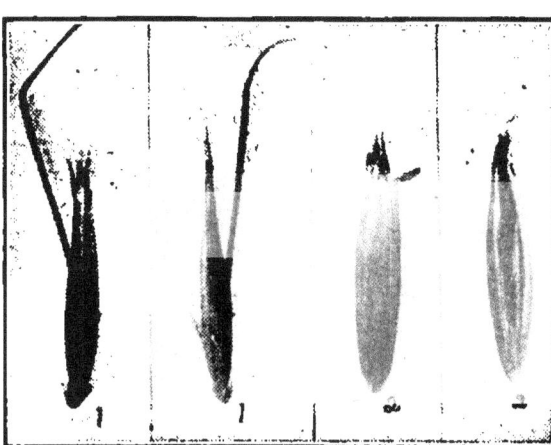

Figure 54.—Seeds of (1) wild oats; (2) tame oats.

long, crooked awn. This awn is not always a safe guide, as it is often broken off in the threshing machine. See Figure 54.

Wild buckwheat is a black, three-sided seed, often found covered with a brown husk. It is nearly the size of a kernel of wheat, and common in grain grown on old fields. See Figure 53.

Wild pea or vetch is a heavy, dark brown or gray seed, round in shape and about the size of, or a little larger than, a kernel of wheat. It closely resembles in shape the common garden pea, and is easily split in halves the same as a pea or a bean. It is common in grain. See Figure 53.

Much more will be learned about the above weeds, if samples of grain are examined and specimens of weeds seed of each variety discussed are found and studied.

Questions:
1. Have you ever seen a farm that was entirely free from weeds?
2. Tell at least two ways in which weeds get into fields.
3. Describe each weed seed you have studied.

Arithmetic:
1. If a farmer sows a 50-acre field of grain with seed containing 3% weed seed, how much land will he sow to weeds? How much will he lose, if his grain yields $15.00 worth of product per acre?
2. If 10% of the crop in a field is weeds, and it requires 4 lbs. of twine per acre, costing 15c. per pound to bind the crop, how much does it cost per acre for twine to tie up the weeds?
3. A farmer has 1,000 bus. of oats threshed; 4 lbs. in each bushel is weed seed. What per cent of his crop is weeds? How many pounds of weed seed has he?

WEED SEEDS COMMON IN GRASS AND CLOVER SEED

Pure Seed.—Sowing grass and clover seed that is not pure is one of the most common ways of getting bad weeds into the land.

Grass seeds are so small that many weed seeds may be mixed with them and not be noticed unless one is perfectly familiar with both the grass seeds and the more common weed seeds.

Where there is a good stand of grass or clover there is very little chance for weeds to grow. Where there is a poor stand—perhaps the result of sowing poor seed, or of sowing on poor soil, or of winter killing—weeds are very likely to spring up and make a good growth. If such a grass crop is cut for seed, the weed seeds are likely to be mixed with the grass seed.

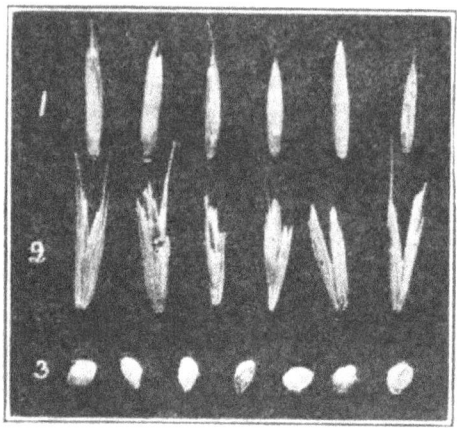

Figure 55.—Seeds of (1) quack grass, single; (2) quack grass with two or more seeds as they grew. They were not separated when shelled. They often appear together. (3) Pigeon grass.

Hay a Cleaning Crop.—If a grass crop is cut for hay, the weeds growing in it are, as a rule, cut before they have had time to ripen seeds. For this reason the hay crop is regarded as a cleaning crop.

Hay with weeds in it is very inferior in quality, and every effort should be made to get such a good stand of grass as to prevent weeds from growing with it. The first step in getting a good stand of grass is to sow good clean seed that will grow. If the grass seed we sow contains weed seeds, we not only sow the undesirable weed seeds, but also sow less grass seed, hence get a poorer stand.

Description.—Below is given a brief description of four kinds of weed seeds most common in grass seed:

Pigeon grass is about one fourth as large as a grain of wheat. It varies in color from nearly light yellow to light green, and has one flat surface. In shape it is similar to half a bean. It is common in grain and in grass seed.

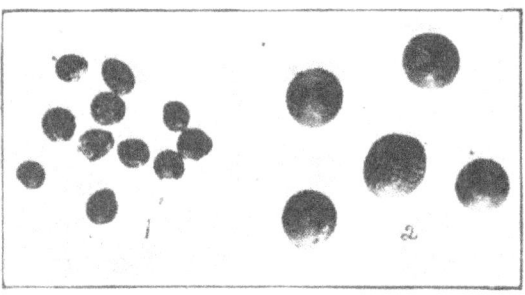

Figure 56.—Seeds of (1) pigweed; (2) wild mustard, enlarged.

Mustard seeds are smaller than a pinhead, almost perfectly round, and dark brown to nearly black in color. They resemble rutabaga seeds and are easily identified by tasting, as they have a sharp, spicy taste. Mustard seed is common both in grain and in grass seed. See Figure 56.

WEEDS AND THEIR ERADICATION

Pigweed.—The seeds of pigweed are small, shiny and black. They are half the size and about the shape of a common pin head. They are commonly found in grain and in grass seed. See Figure 56.

Quack Grass.—Seeds are slender, light in weight, somewhat the shape of oats, but only about one half as long. They are green or light yellow in color. Sometimes two or more seeds are joined together. They may be found in grain or in grass seed, especially in bromus. Quack grass seed is a little heavier, smoother, and more yellowish in color than bromus seed.

We suggest that our readers examine carefully several samples of grass seed found in the neighborhood; first, to become familiar with common grass seeds, as red and alsike clover, timothy, alfalfa, and bromus; and also to learn to identify the weed seeds mentioned above, and to readily observe and know them when seen in a sample of grass seed.

Questions:
1. What is a very common way of getting weeds on a farm?
2. Why is it easier to get bad weed seeds in grass seed than in seed grain?
3. For what reason is a hay crop regarded as a cleaning crop?

Arithmetic:
1. A bushel of timothy seed weighs 45 lbs. What is it worth at 5c. per lb.?
2. A bushel of clover seed weighs 60 lbs. What is it worth at 15c. per lb.?
3. There are 32 quarts in a bushel. Clover seed weighs 60 lbs. per bushel. What does one quart weigh? Timothy seed weighs 45 lbs. per bushel. What does one quart weigh?
4. If a farmer seeded 10 acres of land with grass seed containing 10% weed seeds, how much land would he sow to weeds?

CLASSES OF WEEDS

Habits of Weeds.—If the habits of weeds are studied, it will be found that all weeds may be placed under three classes: annuals, those that live but one year; biennials, those that live two years; and perennials, those that live from year to year.

Annual weeds are those weeds that start from seed, make their full growth, produce seed, and die in one year. In this class we find such common weeds as pigeon grass, lamb's quarter, wild oats, wild barley, mustard, corn cockle, wild buckwheat, French weed, rag weed, etc.

To Eradicate Annual Weeds.—Keep weed seeds out of the soil and prevent the weeds that grow in the field from producing seed. To accomplish this, the following methods will be found useful. Use only clean seed; that is, do not plant the weed seeds. Seed the fields down to tame grass for hay or pasture one or two years in every three to six years. Plant the land to cultivated crops such as corn, or potatoes, once or twice in every three to five or six years,

Figure 57.—A crop of clover hay, a good thing to hold weeds in check.

so that it may be cultivated and the weeds killed in that way. In other words, clean seed and rotation of crops will make it easy to control annual weeds.

Biennial weeds are those that live two years and then die. The first year they start from seed and make part of their growth. They live over winter, then the second year complete their growth, produce seed and die. There are but two common biennial weeds, bull thistle and burdock. We believe every country boy and girl know these two common weeds. These two weeds do not cause trouble in cultivated fields, but are very bothersome in old pastures, along roadsides, and in waste places.

To eradicate biennial weeds it is but necessary to keep them from producing seed. The roots will die in two years, so if no new seed is produced they will disappear. If the land can be plowed and planted to corn or grain for a few years, biennial weeds will disappear, as the plowing each year prevents the plants from getting old enough to produce seed. In pastures, mowing the weeds down close to the ground several times during the summer will prevent them from seeding, and, if this is done for two years, the weeds will disappear. It is difficult to mow these weeds close enough so that they will not produce some seed. The surest method of getting rid of them where the land cannot be plowed is to use a spud, (an implement similar to a chisel with a long handle) with which the plants can be cut off an inch or so below the ground. This is a slow process in a large field, but it is sure.

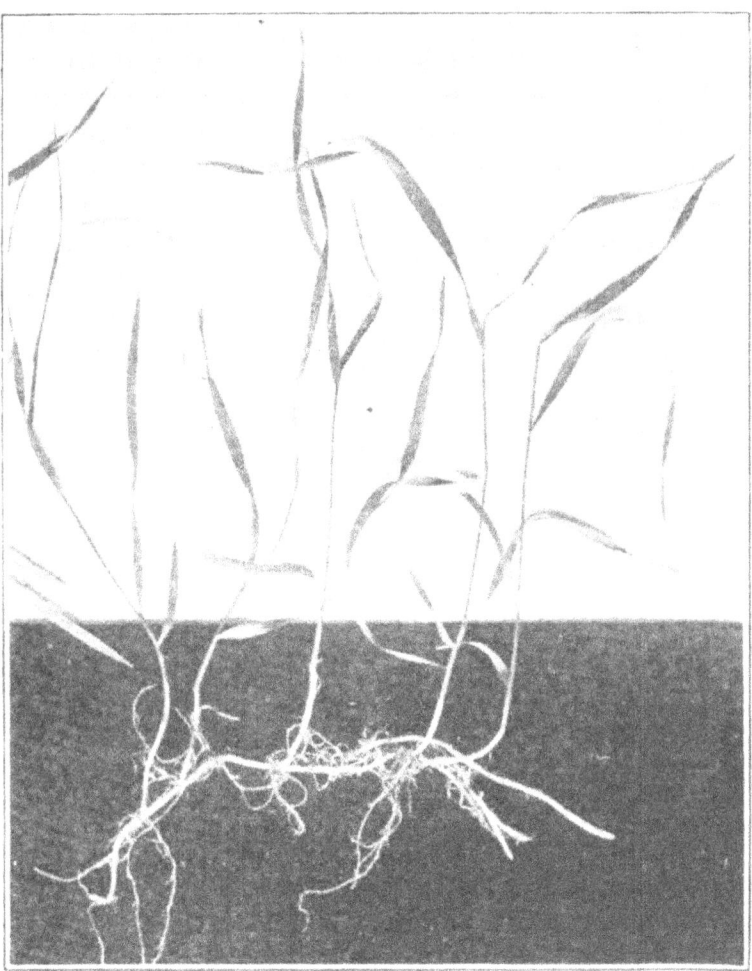

Figure 58.— Roots and stems of quack grass. Note jointed root stalks and that several plants are attached.

Perennial weeds grow year after year, or until something unusual happens to kill them. In this class are found the very worst weeds with which the farmer has to contend. Some of the more common and more troublesome perennial weeds are quack grass, Canada thistle, sow thistle, morning-glory, and curled dock. These weeds not only grow from seed, but persist in growing and spreading even if prevented from seeding. They grow from underground root stalks or stems. When the land is plowed or cultivated, instead of the roots' being killed they are broken into pieces which start to grow and produce new plants.

To eradicate perennial weeds is one of the most difficult jobs a farmer has to do. Many farmers do get rid of them, but some farmers just give up and say it cannot be done. To kill them one must not only prevent the plants from seeding, but must kill the roots as well. The roots can be killed only by digging them out of the ground and removing them from the field, or by starving them out. The roots can be starved only by preventing them from forming leaves. This can be done best by plowing the land very thoroughly, and then by disking and harrowing very carefully and very often for about three months. This requires lots of work, and is very expensive; but, if the work is carefully done, even quack grass can be eradicated.

Bulletins.—For further information about weeds, how to identify them and eradicate them, write to your State Experiment Station and ask for bulletins on weeds.

Questions:
1. Tell what you can about annual weeds. Name some of them, and tell how to eradicate them.
2. Name two common biennial weeds, and tell how to destroy them.
3. Name the worst perennial weeds, and tell how to kill them.

Arithmetic:
1. If a man's time is worth 16c. per hour, and each horse's time is worth 10c. per hour, how much does it cost per day for the labor of a man and four horses? (10 hours for a working day.)
2. If a man and 4 horses can harrow 40 acres per day, how much does it cost per acre to harrow land? (Is not harrowing a cheap way to kill weeds?)
3. Morning-glories twine about and kill 200 hills of corn on an acre. What part of the crop is thus destroyed? (There are 3,240 hills per acre.)

CHAPTER IX

A GARDEN

ITS IMPORTANCE

Value.—A garden is a very small but important part of a farm. After the long winter, during which we have lived largely on bread, meat, canned vegetables and fruits, there is nothing more delightful than to get the fresh vegetables from the garden. They are succulent, easily digested, palatable and nutritious.

A variety of good vegetables means much to every housewife, who must plan and prepare at least one thousand meals during the year. If she has at hand an abundance of fresh vegetables for summer use, and of the same canned for winter, the question of preparing suitable and healthful meals is greatly simplified.

For the boy or girl who wishes to help the mother, and at the same time learn a great deal about soil and how to cultivate it; about plants and how to grow them, there is nothing on the farm that offers a greater opportunity than the garden. A very few minutes of well directed effort will work wonders in the production of many vegetables, such as radishes, onions, etc.

Income from a Garden.—Some boys and girls, living near town, may earn considerable by caring for a few varieties of vegetables and selling the surplus in town, But boys or girls have right at home a good market for as much as they can raise. They may not, and perhaps should not, expect to receive money for what they raise for home use, but they may rightly consider that they earn all that the vegetables would bring, if sold. Vegetables have a value whether sold or used at home. Some farmers seem to think that their living costs nothing. It is true that they do not pay out money for much of their food; but if they would consider, for example, the vegetables used worth what might be received for them, if they were sold, every farmer would see that what is used from a garden amounts to much, and that a garden is a very important part of a farm.

The boys and girls may like to keep an account of the amount and value of garden produce used, to see how much a garden is really worth. Pulling weeds may prove less tedious to a boy who is thinking of how, by his efforts, he is increasing a yield which he is going to record.

An Account with the Garden.—To keep an account of the garden produce, any note book of convenient size may be used. Devote one or more pages to each variety of vegetables. Each boy or girl should consult his parents and agree upon a price for the produce,—such as three cents per dozen for radishes and green onions, one half a cent a head for lettuce, fifteen cents a peck for peas, string beans, etc. On the page of the account book devoted to radishes, record the price agreed upon for radishes; and it might make it more interesting to record also the date when the seed was sown.

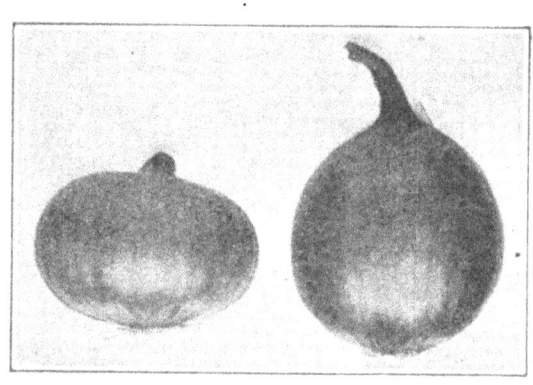

Figure 59.—Flat onion on left, globe on right. Globe onions yield more and usually sell better.

As soon as any variety begins to yield, put down on the page devoted to it the date it was first ready for use and the amount gathered. During the summer each boy or girl will find out at night what vegetables were used that day and the amount of each gathered, either for immediate use or for canning, and record the date and amount.

In the fall, when everything has been gathered from the garden, your note book will show the amount and value of each variety used, sold or stored. The sum of the values of all the varieties of vegetables will be the worth of the garden. It would be interesting, also, to keep a record of any expense for seed or stock and an estimate of the value of any labor spent upon it.

Questions:
1. Why is a garden important?
2. How does a good garden help the mother?
3. How is it of value to a boy or girl?
4. How may we find the value of a garden?

Arithmetic:
1. How many dozen carrots in 3 rows 121 ft. long, if the carrots are 2 in. apart in the row? How much are they worth at 5c. per dozen?
2. What part of an acre is occupied by 3 rows of carrots 121 ft. long, if each row occupies a space 12 in. wide? (There are 43,560 sq. ft. in an acre.)
3. If a boy spends 2 hours each week for 8 weeks in caring for 3 rows of radishes 121 ft. long, how many hours will he spend? How much is his time worth at 10c. per hour?

PLAN AND PREPARATION

Location.—As many trips are made to a garden during the summer, it should be so located that access to it from the house is easy and convenient; and, instead of being located in some little corner where most of the work must be done by hand, it should be accessible from the barn or field, so that most of the work may be done with a horse and cultivator. A good place for a garden is on a south slope sheltered by a grove.

Size.—There is waste land on most farms; and, as long as this is true, there is no excuse for skimping the garden. It should be of sufficient size that room may be given to each variety of vegetable, to permit cultivation with horse labor. A strip about a rod wide at each end of the garden should be seeded to grass, on which to turn when plowing and cultivationg.

Soil.—The soil for a garden should be very rich and productive. More work is required per acre on garden than on field crops; hence the importance of getting good crops to pay for the labor. If grain is sown on soil that will produce but half a crop, six to ten dollars an acre is lost, while, if but half a crop is raised in a garden, owing to the poor condition of the soil, many times as much is lost, because a good garden may yield from $100 to $500 worth of produce per acre.

Preparation of Soil.—Land for the garden should be heavily manured. From twenty to fifty loads per acre may be used. Well rotted manure is best, but other manure will do. It is well to plow in the fall, so that the land will settle down and be less likely to dry out. Fall plowing is also helpful in destroying dangerous insects and worms. Early spring plowing will do; but, in either case, much disking and harrowing should be given the land in the spring, so as to make the soil very fine and mellow before the garden seeds are planted This early harrowing helps to warm up the soil and kills many weeds; also retards the evaporation of moisture. It is a good plan to use a planker or pulverizer to break up all lumps, as securing a fine surface soil makes planting and cultivating much easier.

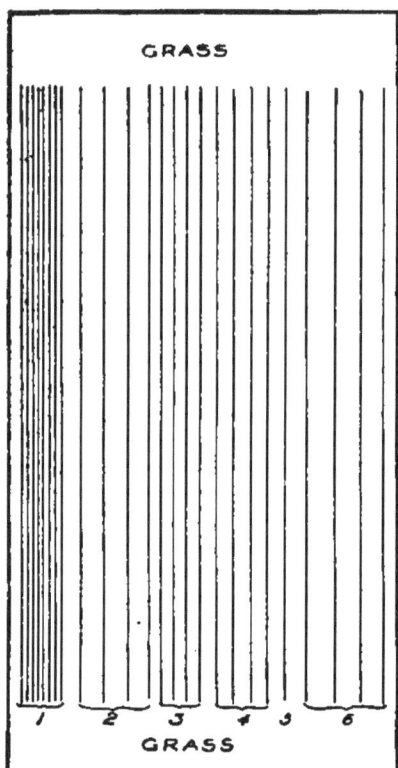

Figure 60.—Garden arranged for convenient cultivation. 1. Rows 14 in. to 18 in. apart for onions lettuce, beets, radishes, turnips, carrots. 2. Rows about six feet apart for cucumbers, melons, etc. 3. Rows three feet apart for sweet corn, potatoes. 4. Rows 4 feet apart for strawberries tomatoes, etc. 5. Asparagus 3 or 4 feet from other plants. 6. Rows about 6. feet apart for berry bushes.

Arrangement.—The rows should extend the long way of the garden. Vegetables which are planted in rows close together (12 to 20 inches) as onions, radishes, carrots, beets, lettuce, turnips, etc., should be on one side of the garden. Cabbage, beans, peas, sweet corn, potatoes, etc., with rows 2½ to 3½ feet apart, should be together. Such running vines as squashes, melons, and cucumbers, which require rows five to seven feet apart, should be together in another part of the garden. This arrangement allows the best use of the horse cultivator. If a row of onions and a row of potatoes were planted side by side, a good cultivation of the potatoes would be likely to cover the onions. This plan also permits the best use of the small hand drill in planting and of the small hand cultivator in cultivating. Everything possible should be done, in the arrangement, planning and preparation of the garden, to reduce the amount of labor required to plant, cultivate and harvest the crop.

Questions:
1. Why should the garden be located near the house?
2. What can you say about the soil and its proper preparation for the garden?
3. What arrangement would you make of the different crops in planting?

Arithmetic:
1. How many cabbages can one raise on an acre, if they are planted 2 ft. apart each way?
2. How many acres of land in a field 10 rods wide and 16 rods long?
3. How many rows of onions, planted with the rows 1 foot apart, can be planted on a field 10 rods wide? If the field is 16 rods long, how long will all the rows of onions be?

SOME COMMON VEGETABLES

Classification.—Vegetables may be classified under three heads, according to their ability to withstand climatic conditions:

(1) Winter vegetables, or those that can be left in the ground all winter without injury, such as rhubarb, asparagus, horse-radish and parsnips;

(2) Hardy spring vegetables, or those that are not injured by frost in the spring, such as onions, radishes, lettuce, cabbage, cauliflower, Swiss chard, peas, rutabagas, carrots, and celery;

(3) Tender vegetables, or those that are killed by frost, such as beans, tomatoes, melons, corn, cucumbers and squash.

Varieties.—There are usually several varieties of each class of vegetables. For example, in tomatoes, there is the Earlianna, Beauty, Ponderosa, Stone, Early Red, Acme, Dwarf Champion, etc.; in peas, the American Wonder, Marrowfat, Telephone, First and Best, Extra Early, etc. It is important for the home gardener to grow only the varieties that one likes personally. In market gardening one's personal tastes should not be considered, but it is necessary to study very carefully the market demand and grow what will sell readily whether it appeals to you or not. A very desirable vegetable for the home garden may be almost valueless in a market garden, because it will not stand handling.

Marketing.—It seems to be quite generally believed that farming consists simply in growing things. This idea does not represent the case. Products must be sold to advantage or there is no profit in growing them. Vegetables are difficult to market, because most garden products deteriorate very rapidly after being harvested. It is, therefore, important that one give very careful study to the selection of varieties, growing products of good quality, harvesting at the proper time and in the proper way, packing them in an attractive, convenient package, and then delivering them promptly and in good condition, when they are wanted.

Figure 61.—Some notable turnips.

Winter Vegetables.—Asparagus, horse-radish and rhubarb are perennials; that is, they grow year after year without being replanted. It is important, therefore, to start them where they are wanted permanently. Roots are usually secured rather than seed, and set out in good rich, mellow soil. If the soil is kept cultivated and well fertilized, it is necessary to harvest only what is wanted. Parsnips are biennial; that is, they live but two years. Seed is planted in the spring early and rather thickly, so as to insure a good stand. The plants are thinned out later, leaving a plant every two to four inches in the row. The roots, the part eaten, may be dug in the fall, and stored, or, if preferred, the roots may be left in the ground all winter and harvested in the spring.

Hardy Spring Vegetables.—Onions are a very important vegetable. They are grown chiefly from seed, planted as early in the spring as the soil may be prepared. The seed is sown in drills from twelve inches to sixteen inches apart.

The soil is kept well cultivated and free from weeds. When the plants are about two inches high, they are thinned to leave about one plant about every two inches. When the tops weaken just above the onion and die down, the onions are pulled, dried, topped and stored. Quite small onions are sometimes pulled and dried and kept over winter, then set out in the spring. They grow quickly and furnish green onions early. They are called sets.

Figure 62.—A field of onions.

Radishes are grown from seed. They require rich, moist soil, and unless grown rapidly are very inferior in quality. The seed is sown early in the spring for an early crop. Several plantings are usually made at intervals of two or three weeks for later use. Radishes should be large enough to use in four to six weeks after the seed is planted.

Peas are sown early in spring; early varieties for early use, later varieties for later use. They are sown in drills from two and a half to four feet apart, depending upon the variety. It is desirable in small gardens to furnish something on which the vines may climb, as they are not strong enough to stand erect alone.

Lettuce, carrots and Swiss chard are grown in about

the same way as radishes. Cabbage and cauliflower are usually transplanted; that is, the seed is sown in boxes in the house or greenhouse for early varieties, or in beds in the garden for later varieties. Then the small plants are later set out in the field where they are to grow.

Celery is not commonly grown, but it should be, and may be grown on any good corn soil. It requires rich, moist soil. The seed is first sown in boxes, or beds, and transplanted. It is often transplanted twice, first from the small beds or boxes to larger beds, then later to the field. For early varieties the seed is sown a month or more before the soil outside can be worked. For late varieties it is sown after the soil outside is in good condition. The plants are set in well prepared, rich soil, in rows three to four feet apart, and the plants about six inches apart in the row. Celery is kept cultivated as any crop. Sometimes a mulch of manure or straw is placed between the rows to check the growth of weeds, and retain moisture. It is necessary to keep soil packed against the plants on either side to keep them from spreading. Celery is always bleached before it is used. This may be done by placing boards on either side of the rows, or by banking up the rows with soil from between the rows. If the celery is to be used in the winter, it is put into the cellar green, then bleached as needed. Bleaching is merely growing celery away from the light. Celery is stored in the cellar by packing the plants tightly in boxes with sand in the bottom into which the roots are placed. It is kept as cool as possible without freezing. To bleach in the cellar, keep the plants moist and warm enough so that they will grow.

Tender Vegetables.—Beans are planted in the spring after all danger of frost is past, in rows two to two and a half feet apart, and with from three to six seeds per foot in the row. Good clean cultivation is all that is required. String beans are secured from varieties planted especially for that purpose. Shelled beans, the navy varieties, are usually planted.

Sweet corn is planted and cultivated the same as field corn, except usually in a much smaller way.

A GARDEN

Tomatoes are grown from seed planted, usually in boxes in the house, in greenhouses, or in sheltered places. When all danger of frost is past, the plants are transplanted to the field or garden. A very few plants will supply the needs of the family.

Melons, cucumbers and squash are very tender and easily killed by frost. They are planted usually in hills from six to nine feet apart each way, on rich well prepared land. They require cultivation and plenty of moisture.

Questions:
1. Name three classes of vegetables and several common vegetables to be found under each class.
2. Describe briefly the methods you would use in growing a crop of onions; a crop of celery.
3. Tell all you can about marketing vegetables.

Arithmetic:
1. If one applies 25 tons of manure per acre, how much does it cost at $2.50 per ton?
2. If an acre of onions yields 600 bushels, what is the value of the crop at 85c. per bushel?
3. How many celery plants can one produce on an acre with rows 3 ft. apart and plants 6 in. apart in the row? (There are 43,560 sq. ft. in an acre.)

CHAPTER X

FRUIT ON THE FARM

ADVANTAGES OF FRUIT

Succulent Food.—That fruit is a healthful form of food is recognized by all, yet there are many persons living on farms who seldom have as much fruit as is necessary for the health and comfort of the family. The need

Figure 63.—An exhibit of fine apples

of succulent food for animals is met by supplying ensilage or roots, yet in many homes the need of succulence in the family diet, though as necessary and easily supplied, is apparently not recognized. Although the juices of fruits help to supply the body with liquid and furnish it with needful mineral substances, these products are especially acceptable and suitable items of diet during the warmer months, when we do not desire so much food of a heat-producing nature.

Home Products.—The home-grown fruits surpass any that can be purchased; for during the summer they may be had clean and fresh, and are very different from the dusty and inferior ones which are often all that the markets have to offer. Such fresh fruit is equally as much better than the market product when canned.

Another great advantage in raising fruit is that the family will then be well supplied, while there are few farm families that will pay the necessary price to secure the amount needed, if it must be purchased.

Every housekeeper knows the value of a good supply of fresh fruit during the summer and fall and the pleasure that is derived from the canned fruit stored away for winter use. The jellies and preserves form a healthful and palatable dessert at any time, and are always ready.

Canned fruit juice, which may be had in large quantities where fruit is plentiful, is perhaps the most appetizing and healthful drink for the sick, and is equally as refreshing for those who are well.

Ease of Supply.—It is neither laborious nor expensive to provide an abundance of both large and small fruits. A small patch of strawberries will produce more than plenty for an ordinary family. Raspberries, gooseberries and currants may be grown at little loss of space along the garden fence. A small orchard, also, will more than suffice for family needs. Ordinary cultivation and occasional spraying are the only demands upon labor. With this attention, however, the results from every point of view will be beneficial and delightful.

Questions:
1. What kind of food do we need in addition to the protein, carbohydrates and fats?
2. What class of foods supply bulk and succulence?
3. What are the advantages of raising fruit for home use?

Arithmetic:
1. If a family of six uses 1 qt. of canned fruit per day, how many quarts will they use in 6 months?
2. If a family of six uses the equivalent of 1½ quarts of fresh

fruit per day, how many quarts would be required to supply them a year? How much are these berries worth at 10c. per quart?

3. If a man spends 5 hours per week for 12 weeks caring for fruit, to produce enough fruit for a family of six as given in example 2, how many hours would he spend? What is his time worth at 15c. per hour?

STRAWBERRIES

Adaptibility.—The strawberry may be grown in almost any locality, from the far north to the extreme south. It is the most widely distributed of the cultivated fruits, and perhaps the most universally popular.

Varieties.—There are several hundred varieties of strawberries listed. Some varieties are particularly adapted to the cooler summers and to the soil conditions of the northern districts, while others are adapted more particularly to the southern conditions. At least one or more varieties may be selected for any district.

There are varieties of strawberries that have imperfect or pistillate flowers, and varieties that have perfect flowers, or flowers containing both stamens and pistils. The perfect may be distinguished from the imperfect only by the flower. When buying plants, one must depend upon the knowledge and honesty of the dealer to secure either plants with perfect flowers or a sufficient number of them to properly fertilize the pistillate flowers. The pistillate varieties are often the best bearers, and are not objectionable when planted with staminate varieties, but are fruitless when planted alone. However, to simplify matters it is wise for the amateur to select the perfect varieties.

Soil.—It is generally conceded that strawberries are most successfully raised on sandy or gravelly loam. New clover sod makes a desirable soil, but it is not safe to use old sod land, on account of the larvae of many injurious insects which are likely to be in the soil and feed upon the young plants. To guard against such, it is well to have the strawberry crop follow some cultivated crop, as potatoes, beans or corn, for the cultivation is quite likely to kill the larvae.

Preparation of the Soil.—The land should be well fertilized. For four rows one hundred feet long, about what would supply the ordinary family, a load of well decomposed stable manure is needed. It is preferable to plow this under in the fall. The surface should be kept pulverized in the spring, until it is ready for the plants.

The Plants.—Strawberries are propagated by runners. The runners grow out from the old plants, and at the joints take root and form new plants. It is these new plants

Figure 64.—A strawberry bed with straw raked between the rows for winter protection. It holds moisture and checks the growth of weeds.

which should be set out. They are distinguished from the old by their white roots. Old plants have dark roots.

The plants, when taken up to be transplanted, should be trimmed of dead leaves or of too large a growth of leaves, and of all pieces of runners. Many roots are desirable; but, as they may make planting difficult, the large roots are usually trimmed.

If the plants purchased seem weak or wilted, or if the field at the time is exceedingly dry, they should not be set immediately. It is much safer to shake them out well and put them close together in a row, where they may be easily protected from the wind and kept well watered. Here they will freshen, perhaps start to grow, and be ready for the field when conditions there are more favorable.

Setting of Plants.—Strawberry plants may be set out at any time, from early spring until midsummer, provided the plants are strong and the ground moist. The earlier they are set, however, the better, as they then have a longer growing season and the roots seem to form more abundantly in the cooler weather of spring.

In setting the plants, the roots should be shaken or spread out as naturally as possible and the dirt firmly packed about them. Care must be taken to set the plants the right depth. The terminal bud should not be hidden, yet the upper portion of the root should be well covered. The safe rule is to set the plants as nearly as possible the depth they were before.

System of Planting.—Strawberries are either set in hills three by three feet apart, or in matted rows. When set in matted rows, the common practice is to set out single rows four feet apart, plants twelve inches apart in the rows. The runners are then allowed to cover a space of from six to nine inches on either side of the plants, making a matted row from twelve to eighteen inches wide and leaving a space of from thirty to thirty-six inches between the rows for cultivation and convenience in harvesting. This space may be reduced, and there is one advantage in the narrower space—i. e., the runners may be allowed to occupy it and the old row be plowed up. This saves resetting.

Cultivation.—Cultivation should be shallow, yet deep enough to destroy weeds and frequent enough to keep the surface well pulverized and to maintain a surface mulch. Moisture is then more readily admitted and evaporation is checked to a considerable extent.

Care must be taken to keep the plants from getting too thick in the row. If too thick, they are less vigorous and produce smaller and poorer berries. When set in rows, enough new plants may be set to make a continuous row 12 to 18 inches wide, with plants not nearer together than 6 inches.

To protect the plants from frequent freezing and thawing, a mulch is applied. It is usually of coarse material, as hay, corn stalks, or straw, and is applied late in autumn or in the early winter. It should protect the plants, yet not smother them. Ordinarily it should be about three inches thick and extend over the entire bed. In the spring most of the mulch is taken from over the row and put in the spaces between, where it preserves moisture and keeps down weeds. The portion left on the row aids in keeping the fruit off the ground.

Questions:
1. What is the best soil for strawberries?
2. How should the soil be prepared?
3. What can you say of the plants to be set?
4. When and how should plants be set?
5. What care should the strawberry bed receive?

Arithmetic:
1. If strawberry plants are set in rows 4 ft. apart and plants 2 ft. apart in the row, how long will four rows need to be to accommodate 200 plants? How much space will they occupy?
2. If strawberries yield 3,000 quarts per acre, how much is the crop worth at 10c. per quart?
3. If strawberries yield 3,000 quarts per acre, how many quarts should 4 rows 100 ft. long and 4 ft. apart yield?

RASPBERRIES

Adaptability.—The raspberry, like the strawberry, has some species which are adapted to almost every locality. One type of the red raspberry extends over a territory from Arizona to Alaska. There seems, however, to be no variety adapted to conditions in the western Dakotas, eastern Montana and Wyoming, or for parts of California, New Mexico and Texas.

Varieties.—The name raspberry, as we use the term to-day, embraces four species of plants, the European, a foreign species, the Native Red, the Black Cap and the Purple Cane, a cross between Black Cap and Red Raspberry.

Soil.—In their wild state, raspberries are frequently found growing upon a variety of soils, but, like other crops, they thrive better and yield more abundantly upon moderately rich soil. The varieties of red raspberries seem to require for best production a richer soil than most of the varieties of black raspberries, the former giving larger yields on moist clay loam and the latter on sandy loam. Preferably raspberries should follow a cultivated crop. Beans, peas, and potatoes are good preparatory crops.

Propagation.—The red raspberries are propagated by

root sprouts. Young succulent plants may be transplanted, if a part of the parent root is taken with them; but one-year-old root sprouts are better. The red raspberry is also propagated by planting pieces of the old root.

The black raspberries are propagated by stolons or layers. To secure new plants, the branches are bent over some time during the summer and their tips covered lightly with earth. They then root quickly. These new plants

Figure 65.—Uncovering raspberries in the spring. They are laid down in the fall and covered with earth to protect them from thawing and freezing.

are left attached to the old plant until the following spring, when the old stem or branch is cut about eight inches above the new roots. The plants are then ready for transplanting.

Setting of Plants.—The red raspberries may be set out during either spring or fall, fall setting perhaps being more generally favored, as sprouts come out very early in the spring and are very liable to be broken off, if transplanting is attempted at this season.

The black raspberries, tip rooters, should be transplanted in the spring, as they are almost sure to winter-kill if disturbed in the fall.

Plants of either kind are usually set two in a hill, hills three feet apart and rows seven feet apart. If the rows extend north and south, the fruit during ripening time will be somewhat shaded by the new growth, which is an advantage. The spacing may be reduced, but wide spacing has some advantages, chief among which is the fact that it admits plenty of sunshine and makes cultivation possible even when the branches are bearing fruit.

When the new plants are set, they are cut off close to the ground, and are not allowed to bear fruit the first year.

The red raspberries, propagated by root sprouts, should be set a little deeper than they were originally. The black raspberries, propagated by stolons or layers, should be set about their original depth.

Cultivation.—Clean cultivation is especially necessary for the red raspberry, as it spreads rapidly if not checked, soon exhausting its vitality. The spaces between the hedges should be plowed at least once a year, and perhaps less injury is done to the roots if spring plowing is practiced. The subsequent cultivating should loosen the soil only to a depth of two or three inches.

Pruning.—Raspberries require summer and winter pruning. The summer pruning consists in stopping the young shoots when they are about eighteen inches high. This tends to produce branches and root sprouts and so increase the wood growth. In the winter cut out all stems that have produced fruit, and dead and diseased ones.

Winter Protection.—In some severe climates raspberries need winter protection. The roots are loosened at one side of each plant, and the top is bent over and covered with earth. A layer of corn fodder or straw may be added later, if more protection seems warranted. This covering should be removed in the spring, and the plants raised as soon as danger of freezing and thawing is past.

Questions:
1. How general is raspberry culture?
2. How many types of raspberries are there?
3. What soil does each type require?
4. How is each type propagated?
5. Tell what you can of time and manner of setting each type?
6. Describe cultivation and pruning of each type.

Arithmetic:

1. If one plants 4 rows of raspberries 100 ft. long, with 2 plants per hill 2 ft. apart in the row, how many plants are required?
2. If one has four 100-ft. rows of raspberries, rows 7 ft. apart, how much space do they occupy?
3. If raspberries yield 2,500 qts. per acre, how many quarts should one get from a patch 28 ft. by 100 ft.?

APPLES

Adaptibility.—By selection and grafting, varieties of apples have been obtained which are very hardy and adapt themselves to wide ranges of territory and vast differences in temperature. Apple-growing is no longer confined to the warmer portions of the South, but is a possibility in the colder sections of the North. Some very choice apples are raised in districts of short summers and cold winters.

Figure 66.—Picking apples. Note care taken to prevent bruising, as bruised apples do not keep well.

Soil.—The chief requirement of soil for apple production seems to be that it contains an abundance of plant food. The poorer the soil, the more careful management becomes necessary. The best condition for apples seems to be a rich, well-drained soil that will retain moisture.

Preparation of the Soil.—Some prefer to raise apples in a sodded tract. Where such is the case, good clover or prairie sod need not be broken up. Holes should be dug large enough to accommodate the tree without crowding its roots. Where cultivation of the orchard is planned the soil should be brought into a tillable condition before trees are set. The latter is undoubtedly the more successful method.

The Trees.—In selecting trees to set, perhaps more attention should be paid to the roots than to the shape of the top. The essentials are a good root system and a thrifty top.

Apple trees do not come true from seed. That is, seed from Ben Davis apples will not produce Ben Davis apple trees. Trees true to variety are secured by grafting branches of trees of desired varieties on roots secured by planting apple seeds. In severe climates it is very essential that these roots be hardy. To be sure of getting suitable stock, it is wise to order trees for planting only from dealers in whom you have confidence. Trees should not be more than four years or less than one year old.

Trees for northern growers should be taken from the nursery in the fall, and kept through the winter in a cool cellar or be buried in trenches in the field.

Setting the Trees.—In sections of severe winters, apple trees should be set in the spring, as they are almost sure to winter-kill, if set in the fall. In sections particularly adapted to apple culture, the trees may be taken from the nursery in the spring and set out. They may also be set in the fall. The spring planting, however, is likely to be more successful than the fall planting.

The depth to set apple trees will vary according to the slope of the land and the quality of the soil. On steep hillsides they must be set deeply enough to prevent the roots from coming to the surface. In rich soil, four or five inches deeper than they were set originally is adequate. Sandy light soil will require deeper setting.

In sections where apple trees grow large, they should be set from thirty-five to forty feet apart. In sections where trees do not attain so large a size, they may be set from twenty-five to thirty feet apart, and trees in one row alternate with those in the next. As a protection against sun scald, trees should lean a little to the southwest.

Cultivation.—If the orchard is to be cultivated, some crop which requires cultivation in early summer but none in the fall, may be grown. Corn and early potatoes are often planted; and in some eastern states, where there is a great demand for tomatoes for the canneries, this crop is profitable, and the cultivation beneficial to the orchard.

Mulching.—There are arguments for and against mulching. Where cultivation is impossible, a mulch of straw, hay or any coarse material, should be maintained. A mulch

Figure 67.—A well kept apple orchard.

of stable manure is beneficial to young trees, but should not touch them.

Where mulching is practiced, it should be renewed when grass begins to grow up through it, and should cover as large or a larger space than the roots are likely to permeate.

Each spring the mulching should be removed, the ground around the tree well spaded, and a mulch replaced.

The chief objections to mulching are that it tends to encourage growth of roots towards the surface, and furnishes a home for injurious insects. These objections may be oversome by removing the mulch in the spring and replacing it in the fall.

Pruning.—The objects in pruning are to direct the growth of the tree, to admit sunlight, and to maintain the vigor and vitality of the tree. Trees well exposed to sun and wind will need less pruning or thinning of branches than trees in more sheltered places. Old, neglected trees may be benefited by pruning, as the remaining branches will receive more of the sap gathered by the roots as the foliage area is reduced. If the trees are badly in need of pruning, it is well to remove only a portion of the surplus branches the first year, and continue the pruning the following years.

A limb that crosses another, or is too near another, should be removed as well as all diseased portions.

Pruning may be done on warm days in early spring. It is safer not to prune when the twigs are frozen, yet the pruning should be done before the sprouts start. A branch or twig should be cut off close to the trunk or the branch, as such cuts heal more quickly than if a stub is left. The scars made should be covered with grafting wax or some similar substance. Pure linseed oil and white lead are successfully used.

Questions:
1. How has apple-growing been made possible in districts not originally adapted to it?
2. What kind of soil do apple trees require?
3. How are different soils or sods prepared for apple trees?
4. Tell what you can of trees suitable for setting.
5. How and when should trees be set?
6. What cultivation is necessary?
7. What mulching and pruning are necessary?

Arithmetic:
1. If trees are set 25 ft. apart each way, how much space does each tree occupy? How many trees can be set on an acre?
2. If trees are set 25 ft. apart each way and each tree produces 4 bus. of apples, how many bushels will be produced per acre? How much are they worth at $1.00 per bushel?
3. If one has 10 apple trees, each producing 4 bus. of apples, how much will the apples be worth at 90c. per bushel?

CHAPTER XI
PLANT DISEASES AND INSECT PESTS

PLANT DISEASES

Prevalence.—Plants are affected by diseases just as animals are, and one of the very serious problems of the farmer is to learn to know the diseases of the crops he grows, and the most practical means of combating them. Many volumes have been written about plant diseases, and there is much that is still unknown about them. Plant diseases are caused by parasitic plants that grow in or on the useful farm crops and steal their living from them. Only a few of the more common diseases of the principal farm crops can be mentioned in a book of this kind.

Loss.—An immense loss is sustained every year from plant diseases. The loss on every farm from this cause is very great, much greater, we believe, than most farmers realize. Every year there is some loss from disease in nearly every crop grown, but as a rule the disease is not noticed unless it is quite bad. For example, one can very seldom, if ever, find a field of wheat entirely free from rust; yet the rust is usually unnoticed until a season occurs which is very favorable to it, and a great loss is sustained. Such a season occurred in Minnesota in 1914, and wheat rust caused a loss of at least 30 per cent in the wheat crop, or a money loss of over $15,000,000 in one state in one year.

Rust affects a great many of the common farm crops, such as wheat, oats, barley, alfalfa, and rye. As indicated above, the loss is often enormous. At present there is no known remedy except to select varieties of crops that are capable of resisting the disease, and selecting early-maturing varieties that are likely to ripen before the rust becomes prevalent enough to seriously injure the crop. Winter grain crops, because they ripen earlier, are less subject to rust than spring-sown crops.

Grain smuts are diseases that affect barley, oats and wheat. The black or naked heads of grain, commonly seen on close examination of a field, are caused by smut. There are several different kinds of smut, all of which are quite common. All the smuts may be controlled by proper treatment. Some of them are easily controlled, while others are more difficult to handle. Every farmer or farm boy may easily know the different smuts and the best means of controlling them. It is very common for smuts to cause a loss of 10 per cent in a grain crop. This means a loss of from $1 to $2 per acre. This loss may be prevented at a cost of only a few cents per acre.

Loose smut of oats, and covered smut of wheat and barley are very easily controlled by treating the seed grain.

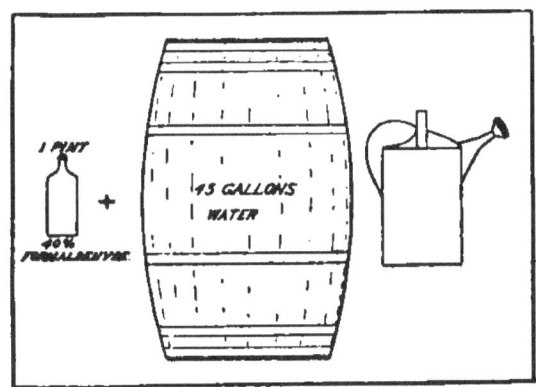

Figure 68. — Material and equipment for spraying.

The smut spores, or seeds of the smut plant, live over winter on the kernels of grain. If the kernels of grain are brought in contact with a solution of formalin, the smut spores are killed. Forty per cent formaldehyde may be purchased at any drug store,—a pint bottle will cost from thirty to sixty cents,—and when mixed with 45 gallons of water will make enough solution to treat 60 or 80 bushels of seed grain. There are machines on the market for treating seed grain for smut, or the grain may be spread on the floor and the solution sprinkled over it with a sprinkling can. The grain must be shoveled over so that all kernels come in contact with the solution. It is well to cover the grain after it is treated with sacks or blankets so that the gas from the formalin will help in killing the smut spores. We give considerable space to these smuts, because they attack very important crops, are very common, almost universal, and are so very easily and simply treated.

Loose smut of wheat and barley causes the whole heads of the plants to fail to produce seed, and instead of the usual head of grain there is nothing left but the bare naked stalk.

A careful examination of a field of wheat or barley after it is well headed out will usually show a considerable number of these naked heads. The spores of this smut are found on the inside of the kernels of grain, and on this account they are hard to destroy. Formalin has no affect on them. The only practical treatment is what is called the hot water treatment and this is very hard to employ. Another method is to get enough clean seed to plant a small plot, and raise the seed grain from that. Write to your State Experiment Station for full information about smuts.

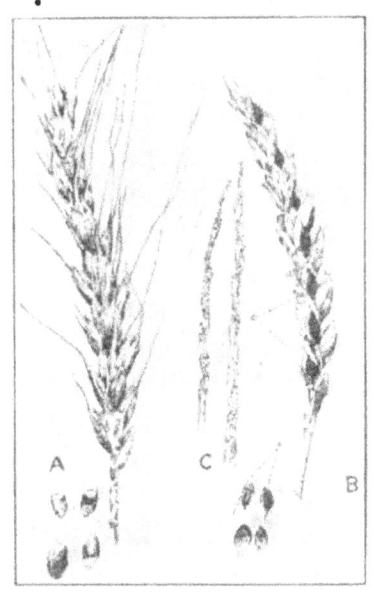

Figure 69.—A is a healthy head of wheat; B, a head affected with stinking smut; C, a head attacked by loose smut.

Corn smut may attack any part of the corn plant. It usually appears in black, soft masses on the ears, stalks or tassels. Each black ball is made up of thousands of spores. These spores live over winter in the soil; so seed treatment is not effective. The only method of control is to pick the smut balls from the corn and burn them, or rotation of crops, or both.

Flax wilt is a disease that affects flax. It attacks the young plants and kills them so they wilt and fall over; hence the name. The spores of this disease live over winter in the soil and on the seed. It is controlled by sowing flax no oftener than once in six or seven years on the same soil, and by using clean seed. If seed from an infested field must be used, it may be treated with formalin the same as oats are treated for smut. In treating flax seed, do not get the seeds too wet or they will stick together and be difficult to sow.

Apple blight, or fire blight, is a disease that attacks the limbs or trunks of apple trees, also other trees. The leaves on the affected parts look as if they had been scorched by fire. The disease also shows on the bark, which turns darker in color. The only remedy is to cut out the parts affected and burn them.

Apple scab is a fungous disease which attacks the blossoms and other parts of the tree, finally manifesting its effect on the fruit, which becomes scabby and shriveled and drops. Extensive losses are suffered in this way. The disease is thought to be more prevalent in wet seasons. The remedy is to spray with Bordeaux mixture once before blossoming,

Figure 70.—Spraying fruit trees.

and again immediately and a few days after the petals fall.

Formula for Bordeaux Mixture

Copper sulphate (blue vitrol), 5 lbs.
Quicklime (not slaked), 5 lbs.
Water, 50 gallons.

Dissolve the copper sulphate in 25 or 35 gallons of water. Slake the lime in the remaining part of the water, and mix. To this amount of Bordeaux mixture ⅓ lb. of Paris green may be added to form a general insecticide to be used on fruit-bearing bushes and trees.

Scab and scale diseases may be similarly treated with the lime-sulphur solution. To control the codling-moth, there should be a midsummer spraying with two pounds of arsenate of lead to fifty gallons of water.

Questions:
1. What can you say of the nature of plant diseases?
2. Learn to write down, without looking, the names of the diseases discussed in this lesson and the plants they attack.

3. Describe the formalin treatment of seed grain, and name the diseases that may be prevented by it.

Arithmetic:

1. If the yield of grain is reduced 10% by smut, how much does a farmer lose who grows 100 acres of wheat with a normal yield of 15 bus. per acre? How many dollars does he lose, if wheat is worth 90c. per bushel?

2. If a farmer pays 50c. for a pint of formaldehyde and uses it and four hours of time to treat 60 bus. of wheat for smut, how much does it cost him per bushel, if his time is worth 15c. per hour?

3. How many acres can one seed at the rate of 1¼ bus. per acre, with 60 bus. of wheat?

DISEASES OF POTATO AND COTTON

Potato diseases are very numerous and cause more loss to the crop than is caused to most other crops by disease. Although the potato crop is very common, and the diseases are very general, comparatively little is widely known or done to prevent the losses.

Potato scab is very common. It attacks the tubers and leaves them with rough unsightly blotches. This scab greatly reduces the value of the crop, though it does not, as a rule, materially reduce the yield. The spores of the disease live over both on the tubers and in the soil. The spores on the tubers are destroyed by soaking the tubers for one and a half hours in a solution of formalin made by mixing one pint of 40 per cent formaldehyde in thirty gallons of water. The seed tubers should be treated before they are cut, then planted on clean land, that is, land that has not produced potatoes for four or five years. Four ounces of corrosive sublimate dissolved in hot water, then mixed with thirty gallons of water, is also effective in destroying scab. The tubers are soaked in it 1½ hours. **Corrosive sublimate is very poisonous and must be handled with care.**

Rhizoctonia is another serious potato disease, which in many sections seriously affects the crop. It affects the vines, the stems and the tubers. It is controlled by treating the seed with corrosive sublimate, as indicated above for scab, and by planting clean seed on clean land.

Potato wilt, or brown rot, is one of the very injurious potato diseases, though it has never been recognized as a disease by a great many potato growers. It affects both the vines and roots, and often causes serious loss in the tubers by causing them to rot. A thin slice cut across the stem end of tubers affected usually shows a dark ring or diseased portion just under the skin. This disease causes a great loss in yield and from tubers, rotting. It may be entirely controlled by selecting clean seed and by rotation of crops. Clip off the stem ends of tubers to be used for seed, and discard all that show the disease, or at least cut off all the diseased part. Select for seed only disease-free tubers, and then treat with corrosive sublimate solution the same as for scab. Plant only clean seed on clean land.

Potato blight causes serious losses in the potato crop nearly every year. There are two distinct blights that affect potatoes. They are called early and late blight. Both diseases affect the leaves of the growing plants, and reduce the crop of tubers by partially or completely destroying the leaves. Late blight also causes the tubers to rot; so it is more serious than early blight. Both of these blights are controlled by spraying the vines with Bordeaux mixture. See page 149. The potato vines must be sprayed from three to five times. Begin spraying when the plants are six to eight inches high, and repeat every ten to fifteen days until the potatoes are ripe.

Cotton Diseases.—There are two serious common diseases of the cotton plant: wilt and root rot. Cotton wilt is somewhat similar to flax wilt. Some plants and some varieties of cotton seem to resist the disease better than others. The selection of resistant varieties and the rotation of crops are the most effective means of control. Root rot affects cotton quite seriously on heavy soils. Deep plowing, drainage and rotation of crops help to control this disease.

Questions:
1. Name some important diseases that affect potatoes.
2. How are the potato blights controlled?
3. Describe the making of Bordeaux mixture?
4. Name two diseases of cotton and remedies for them.

Arithmetic:
1. If an acre of potatoes yields 160 bus., how much is the loss per acre, if the value of each bushel is reduced 5c. on account of scab?
2. If a normal yield of potatoes is 160 bus. and the yield is reduced 25% on account of wilt, what is the loss per acre, if potatoes are worth 40c. per bushel?

3. If five hours of time are required to select and treat enough seed potatoes for an acre, and 30c. worth of material is required for the treatment, what will the total cost of treatment be, if one's time is worth 15c. per hour?

INSECTS AND THEIR CONTROL

Loss to farm crops caused by insects represents a very heavy tax, and increases greatly the cost of producing crops. Many injurious insects, in fact most of them, are always present, and there seems to be little prospect of ever getting rid of them. The problem of the farmer is to know the habits and methods of control of the insects affecting his crops, and to wage continuous war against them.

Habits.—Insects have certain characteristics that distinguish them from other animals. There are three distinct sections to their bodies: head, trunk and abdomen. They also change in character as they develop. There are four quite distinct changes: first, the egg stage; second, the larval or worm stage,—this is the stage at which they do most of the damage to crops; third, the pupal or resting stage, during which time the insect, enclosed in a cocoon, changes to the fourth state, that of the mature insect.

Two Classes.—Insects may be divided into two classes by their methods of eating. Some insects chew their food. These insects can be poisoned by spraying poison on the plants they are eating. Other insects force their sharp mouth parts into the skin or bark of the plant and suck out the juices. Poison sprayed on plants will not affect these insects, because they do not get it. They must be killed by spraying with something that will kill them simply by coming in contact with their bodies, such as soap solution or tobacco extracts. Among the most common biting insects that can be poisoned are potato bugs, cutworms, army worms, currant worms, cabbage worms, grasshoppers, plum curculio, and codling moth. The most common sucking insects are plant lice, squash bugs, chinch bugs, and scale insects.

Remedies.—There are several ways of combating insects, the most generally effective way being good farming; that is, the rotation of crops, fall plowing, clean fence corners, keeping all rubbish picked up about the fields, and maintaining about the place a good grove or other suitable place for birds. Rotation aids in reducing the loss from insects, because it provides for moving each crop to a different field every year or two, and by the time the insects get a start in one field the crop they are attacking is moved to another field and largely escapes. Fall plowing destroys many insects by destroying their winter quarters, or by exposing the eggs or insects so that the weather kills them. Cleaning up fence corners and rubbish destroys good hiding places and winter quarters for many insects. Birds eat great quantities of insects, and anything done to protect or shelter them aids in controlling insects.

Poisons for the biting insects are either sprayed or dusted on the plants. The liquid spray is most common and generally most satisfactory. Paris green mixed in water in the proportion of one pound of Paris green to fifty gallons of water is one of the most common poisons. Arsenate of lead mixed, three pounds to fifty gallons of water, is also generally used. Bordeaux mixture, which is used as a spray to destroy some of the plant diseases, see page 149, may be used with either of these poisons in place of water. So one may often spray for insects, like potato bugs, and plant diseases, like potato blight, at one operation.

Contact sprays for sucking insects may be made at home or purchased. A very simple and satisfactory spray for plant lice is soap solution, made by dissolving one pound of laundry soap in fifteen gallons of boiling water. This solution may be used any time, but is more effective when warm. Tobacco extracts are sold commercially by druggists in serveral different forms. From half a pint to a pint is used in fifty gallons of water.

The cotton boll weevil has done great damage to the cotton crop. When this insect first appeared it was feared it would completely destroy the cotton industry; but it has become less destructive since cotton growers have learned how to control it. Rotation of crops, thorough cultivation of the cotton crop, and early planting of early varieties have been found quite effective in controlling it.

Questions:
1. Can you tell some of the characteristics of insects? What are the four stages or changes that insects go through?

2. Why is a farmer interested in knowing whether an insect gets its food by biting or sucking?

3. What can you say of the cotton boll weevil?

Arithmetic:

1. If corn on fall-plowed land yields 50 bus. per acre and on spring plowed land 25% less on account of the damage done by cutworms, how much is gained per acre by fall-plowing, if corn is worth 50c. per bushel?

2. A sprayed his cabbage with Paris green to kill cabbage worms, and raised 15 tons of cabbage per acre; B did not spray his cabbage and raised only 12 tons per acre. How much did A make by spraying, if cabbage is worth ½c. per pound?

CHAPTER XII

LIVE STOCK

Importance.—Farmers in the United States own almost six billion dollars' worth of live stock, principally as follows: 20,737,000 head of dairy cattle, worth $1,118,487,000; 35,855,000 head of beef cattle worth $1,116,333,000; 20,962,000 head of horses worth $2,291,638,000; 58,935,000 head of hogs worth $612,951,000; 49,719,000 head of sheep worth $200,803,000; and $202,506,372 worth of poultry.

With these vast sums invested in live stock it is very important that the boys and girls who are to handle this vast wealth should know something about animals, their habits and needs. In 1909 the total value of live stock sold and slaughtered and of live stock products sold from the farmers of the United States was nearly three billion dollars.

Live Stock and the Soil.—Aside from the vast investment in live stock and the animal value of live stock products is the great importance of live stock to the farm and to the productivity of the soil. It is a generally recognized fact that farms on which a considerable number of live stock is kept produce larger yields of crops than farms without live stock. This result is true, because, when animals eat a crop of grass or hay or corn, a large part of the plant food contained in the crop is returned to the farm; while on farms without live stock the whole crop is sold from the farm and all the plant food it contained, with it. It is well worth while for anyone who expects to operate a farm to study very carefully the relation of live stock to the farm.

Classes of Stock.—There are several different ways of classifying live stock. First, they may be classified according to the character of the animals, as, Cattle, Horses, Mules, Sheep, Swine, and Poultry. Second, they may be classified on the basis of the uses to which they are put.

Horses and Mules are used chiefly for power or for driving and riding. In some countries in times of famine horse meat is eaten, but such cases are rare. Horse hides or skins are used to make gloves and shoes.

Cattle are kept chiefly for meat and milk. Oxen are sometimes used for work and driving. When animals are slaughtered, their skins are used for making shoes, gloves, harnesses and other leather goods. Their hair is used for plastering. The bones are used for refining sugar, and the bones, blood, hair and other waste is used for fertilizer or for feed for hogs and poultry, and the tallow is used in oleomargarine, for making soap and various kinds of oils and grease.

Hogs are kept for the production of meat and lard, but the waste products are used for fertilizers.

Sheep are kept for their wool from which clothing is made and for meat.

Poultry is kept for meat and eggs.

Questions:
1. Tell what you can about the importance of live stock in the United States. In your community.
2. Why do live stock farms usually produce larger yields of grain and corn than farms without live stock?
3. What can you say of the uses of the different kinds of live stock?

Arithmetic:
1. Figure as nearly as you can the value of all of the live stock on your home farm, or on some other farm where you know conditions.
2. Ten bushels of corn contains 10 lbs. of nitrogen 1.7 lbs. of phosphorus, and 1.9 lbs. of potash. If 10 bushels of corn will produce 100 lbs. of pork which contains 1.8 lbs. of nitrogen, .3 lb. of phosphorus and .1 lb. of potash, what is the value of the fertility saved by feeding the 10 bushels of corn and selling the 100 lbs. of pork? If nitrogen is worth 18 cents per lb. and phosphorus and potassium are worth 6 cents per lb.

CARE AND MANAGEMENT
GENERAL

Chores.—Farm boys are, as a rule, occupied a considerable portion of the time, mornings and evenings, caring for the stock. As the profits derived from the live stock depend to a great extent on the care they receive, it is worth while to spend a little time considering how the "chores" may be done more quickly, more easily and better.

The amount of live stock kept on the average farm must increase as more intensive systems of farming become necessary. The proper care and management of live stock is a matter of increasing importance.

System.—System in doing the chores is fully as effective as system in studying. A carefully thought out plan that will enable one to get a certain amount of work done with the fewest steps and least possible delay will often change

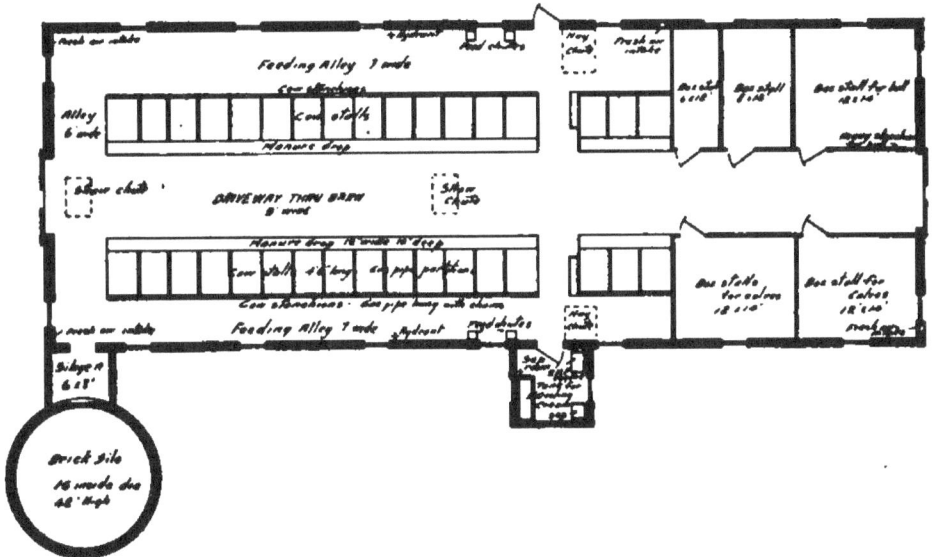

Figure 71.—A well arranged barn in which the "chores" may be done in a minimum amount of time.

a tedious and unpleasant chore time into pleasant pastime. To lead four or six horses to water, and then clean the stable with them in their stalls, requires much more time than is necessary, if yards and watering trough are so arranged that the horses can be turned out to drink by themselves, and the stable cleaned while they are out. If horses are turned out to drink before they are fed their grain, as they should be, they will, as a rule, return to the barn promptly. If you are not caring for your horses in this manner, do so and notice the saving of time; or if you are doing so, notice the time it takes to care for your horses, and compare notes with some neighbor who is still leading his horses out to water.

Value of Time.—Many men who work in factories and at other occupations are paid by the hour on the basis of the amount of work they can do, and every minute of their time must be made to count. Such rules are not possible on a farm, but a young man who expects to farm can and should make himself just as proficient as possible, not with the idea of learning to do twice as much in a day as an ordinary man can do, but to so direct his efforts and utilize his time as to be able to do a good day's work as easily

Figure 72.—Hereford calves out for exercise in a protected yard. Stock should be left out in winter only so long as comfortable.

and in as short a time as is consistent with good work.

Some men have a system of harnessing and hitching a team, and can do these things better and in much less time than the man who has no system and, consequently, does them in a different way each time.

Milking ten or twelve cows twice each day is a comparatively easy task for a man who can milk them in an hour, while to milk the same cows would be almost drudgery to the man who can milk but five or six in an hour. Men can make themselves very proficient and learn to do things

LIVE STOCK

rapidly and well by application and practice, as is illustrated in corn husking. Twenty-five bushels of corn is a fair day's husking for a beginner, but many men by practice can husk from sixty to over one hundred bushels in a day.

Make Animals Comfortable.—One of the first essentials in caring for animals is to make them comfortable. Hogs cannot fatten, hens cannot lay, cows cannot give a good flow of milk and horses cannot continue to do a good full day's work, unless they are made comfortable. Every time an animal is made uncomfortable, either by being left hungry, thirsty or cold, by lying on a hard bed or by being dogged or pounded, the owner losses money by getting less returns from the animal.

Exercise.—Exercise is of the same value to animals as to men. It gives firmness and strength of muscle, promotes good circulation and improves the tone of the vital organs. These conditions all tend to increase the vigor and productivity of live stock.

Outside Feeding.—For those animals that are fed in winter outside the stable proper feed boxes, forage racks and water troughs should be constructed. The first two should be situated so that feed can be hauled close to them, and around all gravel should be spread or a plank platform made. Water ought not to be allowed to freeze. If tanks are used, they may be surrounded with a framework and covered with lids. The space between the frame and the tank may be filled in with sawdust or manure. Care in these particulars is well worth while.

Questions:
1. Why should we plan to save time?
2. What is the most important thing in caring for animals?
3. What is the value of exercise to live stock?
4. Describe outside arrangements for feeding.

Arithmetic:
1. A man by having a handy barn may save 20 minutes per day in doing chores. How many hours may he save in a year? How much is this time worth at 18c. per hour?
2. It requires A 60 minutes to milk 5 cows. If by application and practice he learns to milk 5 cows in 40 minutes, how much time is saved in a week? In a year?
3. From the above, at 20c. an hour what is the time saved in one year worth? How much in 5 years?

160 ELEMENTS OF FARM PRACTICE

SHELTER

Shelter.—To make live stock comfortable in the northern part of the United States and Canada, good shelter must be provided. Expensive shelter is not necessary, but buildings should be so constructed as to keep the animals warm. If they are not kept warm by shelter, some of the food they eat will be used to warm them, and it is cheaper to provide good shelter. Besides, if animals are not comfortable, they cannot do well. This is especially true of milch cows and young stock. Steers do not require very warm quarters so long as they are protected from the storms and the wind.

Light.—Plenty of windows should be provided, so that the sunlight can reach just as much of the interior of the building as possible. Sunlight is a deadly enemy of bacteria and disease germs. There is no better disinfectant than sunlight, and it is so cheap that every building should be amply supplied. Tuberculosis is very common among domestic animals, and it is generally believed that it may be transmitted from animals to man, especially in milk. So it is not alone for the comfort of the animals and the profit we derive, that we provide healthful quarters, but to guard the health of the family as well.

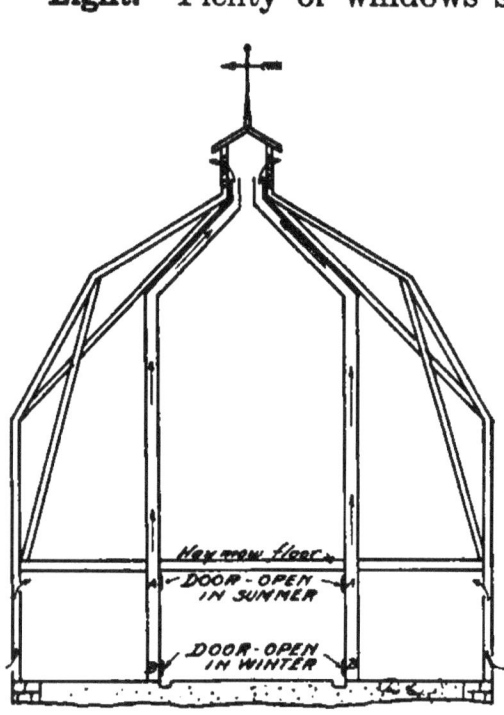

Figure 73.—A good system of ventilation for a barn. Fresh air comes in near the ceiling. Foul air is taken out from near the floor. After King.

Ventilation.—By good ventilation we mean such a system as will remove the foul air from the stable. Leaving a door or a window open is not good ventilation, as it causes a draught and lets out the warm air. A better way is to have one or more flues built in the barn, that will carry the foul air out. Warm air is lighter than cold air, hence it rises. A stable is warmer near the ceiling, if the ceiling is tight, than near the floor. Hence the ventilator flue should open near the floor so as not to take out the warm air. The carbon dioxide exhaled by animals is heavier than air, hence it settles to the floor and will be taken out by such a ventilator. Air should be let into the stable near the ceiling, whence it will gradually settle and become partially warmed before it reaches the animals. The illustration in Figure 73 shows how a ventilator should work. Notice the barns in the neighborhood and make a note of the number of windows and ventilators. How is your barn lighted and ventilated?

Conveniences.—Since chores are so frequent, just a few minutes lost each time they are done amounts to many hours in a year, probably several days; hence it is well to plan to have the barns handy. Feed, both hay and grain, should be stored close to the feeding alleys, which should be sufficiently large. Cleaning the stables is probably the heaviest part of the chores, and should be made as easy as possible. If stalls are so arranged that a team can be driven through the barn and the manure loaded in a spreader, wagon or sled, and hauled directly to the field, it will be much easier than where it is necessary to throw it to one end or side of the barn, then throw it out of a door or a window and pitch it into a wagon outside. This last method is common on many farms, and results in a great loss of labor and time, besides a loss in the value of the manure. The sooner manure can be put on the field after it is taken from the stable the better. A manure pile lying under the eaves of a barn for a few months may lose one half of its value. If a new barn is being built, or the old one changed, a great deal of thought and study should be put on it to make it as convenient as possible. A day spent in planning the alleys, stalls, etc., may save many days' work each year.

Cost.—A feature that must not be overlooked is the cost of a barn. A barn is built to shelter stock, and stock is kept for profit. When a barn is built, it must earn for the farmer enough each year to pay interest on the investment, cost of insurance, taxes, repairs and yearly depreciation. If a barn costs $1,000 and lasts 25 years, the yearly depreciation is $40. Good barns are desirable; but sometimes barns are built on farms where live stock is so poorly managed that it fails to bring in even the yearly cost of the building; and in such cases the investment results in a loss. Care should be taken that the yearly cost of shelter is not so high as to take all profit from the animals kept. By careful attention to the conditions given above, one will get an idea of how to solve such problems.

Questions:

1. Why should buildings be warm? Why light? Why well ventilated?
2. Why should buildings be made convenient?
3. Is there any danger of investing too much money in farm buildings?

Arithmetic:

1. (a) How much is the interest charge per year on $1,000 at 4 per cent?

(b) How much is the yearly cost of insurance on the barn at 50c. per $100?

(c) Such a barn will cost about $10 per year for repairs. What is the total yearly cost of the barn?

(Note: It is the sum of the yearly depreciation, interest, insurance and repair charges.)

2. If such a barn will shelter 20 head of stock, what will be the yearly cost per head for shelter?

3. In a similar manner figure the yearly cost per head for shelter in a $4,000 barn that will shelter 40 head of stock. (Note: Find depreciation, interest, insurance and repairs at same rate as in example No. 1.)

CHAPTER XIII
FEEDS AND FEEDING

Source.—We have learned that the carbon dioxide exhaled by animals is used by growing plants; that this carbon dioxide in the air unites with water and other elements taken by the roots of the plants from the soil, and forms starch and other compounds of which plants are composed. The heat or energy given off by the sun is used to build up these compounds in the plant. Animals are dependent then, upon plants for all their feed.

Requirement.—The act of living involves energy and the necessary consumption of nourishment, as a fuel, to supply the vital force. As the heart works to send supplies to the various parts of the body, it must itself undergo a constant repair, and hence it has its own blood vessels. To think and to walk are at the expense of energy created by the consumption of parts of the body. To run requires more energy and, therefore, causes more waste than to walk. So, then, the harder animals work the more nourishment, or feed, they must have.

How Made Available.—When plants, such as grass, hay or grain, are eaten by animals and digested, the compounds they contain are broken down and used by the animal body, and the energy required to build up the compounds in the plants furnishes energy to the animal.

Selection.—By chemical analysis men have been able to determine the exact constituent elements of plants, that is, how much hydrogen, oxygen, nitrogen, carbon, phosphorus, etc., has entered into their composition. They know, also, the proportion of these elements that is found in animal bodies. With these facts before us, therefore, it is easy to select the proper feeds for each kind of animal according to its composition and peculiar physical adaptability to get the most benefit from this or that kind of feed. Different kinds of animals, and animals doing different classes and amounts of work, require different kinds and amounts of feed. So the intelligent feeding of animals requires a knowledge of the composition of different feeds and of the requirements of the various classes of animals doing different kinds and amounts of work.

Kinds.—Feeds are of two general classes, roughage and concentrates. Roughage includes all bulky feeds as hay, fodder, straw and silage. Concentrates include all the grains and mill feeds, such as oats, bran, corn and oil meal.

Composition.—Feeds are generally divided, according to the elements they contain, into the following classes:

Protein is a term applied to a group of compounds containing nitrogen. Protein is used by animals to make muscle or lean flesh, bone, hair or wool, tendons, nerves, casein and albumin in milk, etc. No other compound can take the place of protein; consequently it is very important that enough be fed, or the animal cannot keep up in flesh and production or work. If too much protein is fed, it can replace the other elements; but as these feeds are usually expensive, it is not wise to feed more than is needed.

Feeds containing a large proportion of protein, as bran, oil meal, clover, etc., are called nitrogenous feeds.

Carbohydrates are those substances in feed that are composed of carbon, hydrogen and oxygen, but have no nitrogen. Sugar, starch, fibre, etc., are carbohydrates. They are used in the body to produce fat, or are burned in the body to produce heat or energy. They cannot be used in place of protein.

Fat.—The oils, wax and fats contained in feeds are called fat. It is used in the animal body for the same purpose as carbohydrates. One pound of fat is worth as much as 2.2 pounds of carbohydrates.

Feeds containing a large proportion of carbohydrates and fat, as timothy or wild hay, corn, barley, etc., are called non-nitrogenous feeds.

Ash.—Plants contain some mineral matter, a small amount of which is necessary in animal bodies, as iron in the blood and calcium in the bones

All the common feeds contain protein, carbohydrates, fat and ash, and hence might be classed in different groups. Those containing a proportionately large amount of protein are called *nitrogenous*, and those containing proportionately large amounts of carbohydrates and fat, *non-nitrogenous*. Sometimes feeds are classed as grains or *concentrates* and *roughage*. The following table shows the composition or digestible nutrients of common feeds:

Digestible Nutrients in One Pound of Feed

(1) Nitrogenous Grain Feeds.

	Protein	Carbohydrates	Fat
Oats	.107	.50	.038
Shorts	.13	.46	.045
Bran	.119	.42	.025
Oil meal	.302	.32	.069

(2) Non-Nitrogenous Grain Feeds.

	Protein	Carbohydrates	Fat
Corn	.079	.67	.043
Barley	.084	.65	.016
Emmer (Speltz)	.10	.70	.021
Rye	.095	.694	.012

(3) Nitrogenous Roughage.

	Protein	Carbohydrates	Fat
Alfalfa	.117	.41	.012
Red Clover	.071	.38	.018
Mangels	.010	.05	.002

(4) Non-Nitrogenous Roughage.

	Protein	Carbohydrates	Fat
Corn Silage	.012	.14	.007
Corn Stover	.014	.31	.007
Corn Fodder	.037	.41	.015
Slough Hay	.026	.42	.011
Timothy Hay	.028	.43	.014
Prairie Hay	.030	.42	.014
Oat Straw	.013	.39	.008

Balanced Ration.—Very few plants contain, in the right proportion, the elements required by animals; so a combination of two or more kinds of feed is found necessary to supply the animal with needed feed. When one gives just the right kinds and amounts of feed to supply the needs of the animal, he is feeding a balanced ration. Such a ration is most satisfactory and economical, because it supplies all the elements needed, with no surplus of any one. All animals require feed that contains in the proper proportion, protein, carbohydrates, fat and ash.

Balanced rations will be given for each class of animals as it is discussed.

Feeds Compared.—In comparing the grain feeds or concentrates with the roughage, one will see that a pound of grain usually contains more nutrients than a pound of roughage. As a rule there is much unsalable roughage on the farm, while grain is salable; hence there is a tendency to feed more roughage and less grain. A reasonable amount of roughage is desirable, but an animal that is working cannot eat and digest enough of it to supply its needs. It should have some grain. To feed only grain is not desirable, as it is expensive and does not supply enough bulk. A proper balance must be maintained.

The feeding of different kinds of animals will be discussed in the chapters treating on those animals.

Questions:
1. What is a "balanced ration"?
2. What is protein, and for what is it used in the animal body?
3. What are carbohydrates, and for what are they used in the animal body?
4. What substances are known as fat, and for what are they used in the animal body?
5. What classes of feeds are called nitrogenous feeds? What classes are non-nitrogenous?
6. Why do animals need food?

Arithmetic:
1. Bran, oil meal, and clover are feeds rich in protein. How much is each worth per pound when oil meal is worth $35.00 per ton, bran $24.00 per ton, and clover hay $5.00 per ton?
2. Corn, barley and timothy hay are feeds rich in carbohydrates. How much is each worth per pound when corn is worth 54c. per bushel (56 lbs.), barley 56c. per bushel (48 lbs.), and timothy hay $5.00 per ton?
3. There is 7.1% of protein in clover hay. How many pounds of protein in one ton? How much does the protein cost per pound, if clover hay is worth $5.00 per ton?
4. There is 11.9% of protein in bran. How many pounds of protein in one ton? How much does the protein cost per pound, if bran is worth $24.00 per ton?

CHAPTER XIV

HORSES

TYPES AND BREEDS

Breeds.—There are many different breeds of horses, just as there are many breeds of cattle. People in different countries or communities have wanted horses for different purposes, and have kept them under different conditions

Figure 74.—A Percheron stallion, a draft type.

until they have developed distinct breeds. We are told that our many different breeds of horses have all been developed from the prehistoric horse, skeletons of which have been found in different parts of the world.

Types of Horses.—There are four general types of horses, ponies, light horses, coach horses, and draft horses. In

Figure 75.—A Clydesdale stallion, a draft type.

each class there are a number of breeds. By far the most important type is the draft horse.

Draft horses are heavy, slow-moving animals, used to do most of the heavy work in the world, such as plowing, grading and hauling. The draft horse has a broad back, broad chest, deep body, relatively short legs, big feet and has great strength, but cannot travel fast. Draft horses are expected to weigh from 1,500 pounds to over 2,000 pounds. Percheron horses originated in France, but are the most common and most popular draft horse in this country. They are black or gray in color and normally weigh from 1,800 to 2,300 pounds. French draft horses are quite similar to Percheron. Clydesdale horses originated in Scotland, but are very common in this country. They are about the same weight as the Percherons. They are usually light bay in color and have one or more white feet and a white stripe or star in the face. They have long

Figure 76.—A Belgian stallion, a draft type.

and shaggy hair on their legs. They may be brown, black, gray or chestnut. Shire horses originated in England. They are about the same in size as the Clydesdale and Percheron, possibly a little heavier and coarser. They are colored like the Clydesdale and have long hair on their legs, but they are more like the Percheron in form. They are not so common in this country as the Percherons and Clydesdales.

Belgian horses originated in Belgium. They are probably a little lighter in weight than the other draft breeds mentioned, and are a little more blocky in form, and have shorter legs. They may be bay, brown or roan in color. They have no long hair on their legs. This breed is not common in America. Sulfolk horses are another English breed not common in America, but well worthy of consideration. They are not quite so heavy as Percherons, are chestnut in color, and have no long hair on their legs.

Figure 77.—A good type of light driving horse.

Light horses, as indicated by the name, are light in weight, weighing from 800 lbs. to 1,200 lbs., or sometimes as much as 1,300 lbs. They are used for driving, riding and racing. These horses are distinguishable by their light bodies, longer, finer legs, and more graceful, easier action than is seen in the draft horses. Several more or less distinct breeds are found in this class. The thoroughbred was developed in England and is used for running and hunting. The Arabian is a beautiful horse originated in Arabia, and used for riding. The Orloff Trotter originated in Russia and is used for driving and racing. There are three American breeds in this class, American Saddler, Standard bred and Morgan, used for driving and racing.

Coach horses include several breeds of horses used for hauling heavy carriages or coaches, and sometimes for light hauling. They are lighter and have much better action than draft horses and are heavier and slower than the breeds mentioned under light horses. The breeds in this class range in weight from 1,000 lbs to 1,500 lbs. Hackneys and Cleveland Bays are English breeds. Hackneys are used extensively as fancy carriage horses. Cleveland Bays are used for heavy coaches and for light hauling. French coach horses, originated in France, and German coach horses, developed in Germany, are other important breeds coming under this class.

Ponies include several breeds of small horses used for driving and riding, chiefly by children. This class includes Shetland ponies, Indian and Mexican ponies, etc.

Questions:
1. What is the most common breed of horses in your community?
2. Describe fully the breed of horses you like best.
3. Name and describe all the common breeds of draft horses.

Arithmetic:
1. A is plowing with two horses, weighing 1,000 lbs. each. B is plowing with five horses, weighing 1,500 lbs. each. What is the weight of each man's team?
2. If a man with two 1,000-lb. horses can plow 4 inches deep with one 14-inch plow, how deep can B, in Example 1, plow with two 14-inch plows (a gang plow), assuming that the depth B can plow will be in proportion to the weight of his team.
3. If A and B (see example 1) each feeds his team at the rate of 1¼ lbs. of oats per day, per 100 lbs. live weight, how much would each have to pay for oats at 40c per bushel?

CARE AND MANAGEMENT

In the Stable.—While in the stable the horse should have plenty of good bedding to keep him comfortable and clean whenever he wishes to lie down. Good ventilation is very necessary, also, as he requires about one cubic foot of fresh air every hour for each pound of his weight. The feed box should be broad and flat-bottomed, and kept clean. He should be well tied so as to permit his head to rest on the floor and yet the strap should not be so long that he can put his foot over it.

Grooming.—Horses should be thoroughly curried and brushed every day for the sake of both health and appearance. Washing may be desirable at times, but is likely to spoil the lustre of the hair and may cause chill unless promptly dried. The best time to groom the horses is in the evening after the work is done.

Shoeing.—Horses at work need to be shod to prevent the hoof from wearing away and becoming sore, and, in winter, to prevent their slipping. Careful attention to shoeing adds much to the comfort and safety of a horse as well as to the convenience of the owner.

The cost of horse labor has been determined by the Minnesota Experiment Station, in co-operation with the U. S. Department of Agriculture, by keeping accurate records on twenty-four Minnesota farms for six years. These records show that the total cost of keeping a farm work-horse for a year averages about $84.00, and that the average number of hours of work done by each horse is about 1,000 per year, making the cost per hour of work 8.4 cents.

To one who has given the matter little thought, the above figures seem high, but, when one considers that the cost of a horse for a year includes several items, it becomes plain that the figures are not far wrong. The following items are the important ones in the cost of keeping a horse: feed, labor for caring for him, depreciation, interest on investment, shelter, shoeing, and depreciation and repair of harnesses. While it is not easy to figure all these items without having kept accurate records for a considerable time, yet a fairly accurate estimate may be made by noting about the amount of hay and grain fed per day in winter and in summer, and its value, the amount of time spent each day in caring for the horses, the value of the horses and on this value figuring the interest and depreciation, and then adding to these items a fairly liberal estimate of the cost of shelter, harnesses and shoeing. The total amount will no doubt be a surprise.

The cost of horse labor on the farm may be reduced by raising more horses on the farm, by keeping fewer work horses, by economical feeding and by better planning of the work.

Raising More Horses.—By raising more horses on the farm, most of the farm work may be done with mares that raise colts, and with young horses before they are old enough to sell to advantage. In this way the horses will be doing double duty. The item of depreciation will be largely eliminated; also a part of the cost of feed, because at least a portion of the feed fed to the mares will be paid for out of the value of the colts.

Fewer horses can often be kept, with little inconvenience in doing the farm work. When one realizes that it costs $70 to $90 per year to keep a horse, he may find that often an extra horse may be hired for a few days, during seeding or harvest, cheaper than to keep one a whole year when it is really needed but a very short time.

Distribution of Horse Labor.—One can often greatly reduce the demand for horse labor at special seasons, and distribute it over a longer season, by following a diversified system of farming. If a part of the farm is seeded to grass, there is less plowing and seeding to be done. If corn is raised on a part of the land, instead of seeding it all to grain, the seeding and harvesting periods are made longer; so the same land can be handled with less horse labor.

Have Plenty of Horses.—It is important, however, to have plenty of good, strong horses with which to do the farm work, because with good horses one can do more work in a day than with poor ones, and thus save man labor, which is also costly. It is easier to keep good hired men, if one has good horses. The good farm manager will consider the question from both sides and act accordingly, but he cannot act wisely unless he knows all the facts.

Questions:
1. Name some of the items that must be considered in determining the cost of keeping a farm work horse.
2. Give four ways by which the cost of horse labor may be reduced on the farm.
3. Give at least two reasons why plenty of good horses should be kept on the farm.
4. Why should horses be well bedded? Groomed? Shod?

Arithmetic:
1. If a horse is fed 2 qts. of oats three times a day half the year and 4 qts., three times per day for the remainder of the year, how many bushels of oats will he be fed in a year? How much are they worth at 35c. per bushel?
2. If a horse is fed 15 lbs. of hay per day, how many tons will he be fed in a year? How much is it worth at $6.00 per ton?

3. If a man spends 1 hour per day caring for horses, how much time will he spend in a year? How much time is this per horse? How much is this labor worth at 14c. per hour?

FEEDING

Feed is the greatest general expense in keeping a horse. This fact and the fact that there are many different ways of feeding and kinds of feed, make it evident that in the feeding of horses is a great opportunity for waste or saving. The average cost of feed (hay or grain) for a farm work horse has been found by the Minnesota Experiment Station to be about $50.00 per year. If one is keeping several horses, there is an expenditure sufficiently large to be worthy of consideration.

Concentrated Food.—A horse has but one stomach and that is not nearly so large as the stomach of a cow. On this account a horse can not eat as much roughage (hay and fodder) as a cow, and must have a larger proportion of concentrated feed, as corn, oats, etc., and horses have less time in which to eat than cattle.

Figure 78.—Horses at work.

A horse at work is busy for about ten hours each day, and has only the night in which to eat less concentrated feeds. Four pounds of oats, which is a fairly good feed for a horse, contains .42 lbs. of protein, 2.0 lbs. of carbohydrates and .15 lbs. of fat. Four pounds of timothy hay contains .11 lbs. of protein, 1.72 lbs. of carbohydrates and .06 lbs. of fat. A horse can eat 4 lbs. of oats in ten to twenty minutes, while it would take him about an hour to eat four pounds of hay.

At noon a work horse seldom has a chance to eat as much as he wants; but, if he has a good feed of grain, he can come more nearly getting what he needs than if he is fed only hay.

More Grain Than Roughage.—For the reasons given above, a good ration for a horse at heavy work must contain more grain than roughage by weight; while a good ration for a milch cow will contain fully twice as much roughage as grain.

Roughage is a much cheaper form of feed than grain. For example, oats at 32c. per bushel cost 1c. per pound; while hay at $5.00 per ton costs but 1c. for four pounds.

Since a horse at work must have a large proportion of grain, it is very important that one try to feed as cheap a grain as possible to get the desired results.

Oats and Corn.—The favorite feed for horses is oats and timothy hay. The average cost of oats per pound is over 1c., while the average cost of shelled corn is about ¾c. per pound. The grain feed alone for a horse costs about $50.00 per year. If corn can be used to replace all, or a part of, the oats, a considerable saving can be made.

We have learned that there are two general classes of feed: nitrogenous or muscle-forming feeds, and non-nitrogenous or fat and heat-forming feeds. Most of the common farm feeds have both nitrogenous matter (protein) and non-nitrogenous matter (carbohydrates), but the proportion of these compounds varies.

The most common feeds that have a comparatively large amount of protein, are bran, oats, clover and alfalfa hay. The feeds containing large quantities of carbohydrates are corn, barley, timothy or wild hay, and corn fodder. In the grain feeds, those containing a large proportion of protein are more expensive than those rich in carbohydrates; while in the roughage, clover hay, which is rich in protein, is cheaper than timothy, which is very poor in protein. These facts make it evident that the needed protein may be furnished in roughage cheaper than in grain.

Corn and Clover Hay.—If corn is to replace oats as the grain feed for horses, some feed rich in protein must be used for roughage. Clover is the cheapest form of roughage that can be produced on the farm, and is rich in protein. There is a general belief among horsemen that clover hay is not good for horses. Poor and dusty clover hay may not be; but good, bright, well cured clover hay, fed in moderation, is a very good kind of hay for horses.

Figure 79.—Making clover hay. Clover should be grown and fed on every farm. It is a very cheap nitrogenous feed.

Recent experiments at the Ohio Experiment Station have shown that horses fed mixed timothy and clover hay, kept just as well, were able to do just as much work and showed just as good spirit, when fed corn as when fed oats, and that a pound of corn on the cob was worth as much for horse feed as a pound of oats. If horses can be fed corn and clover hay without detriment to them, the cost of keeping a work horse can be reduced $10.00 to $20.00 per year. These data are worthy of study and a fair trial.

Questions:
1. Can you give any way by which the cost of feed for a horse may be reduced without injury to the horse?
2. What can you say of the relative value of corn and oats as feed for horses, and the cost per pound of each?
3. Compare timothy and clover hay as to their feeding value.

Arithmetic:
1. If corn is worth 54c. per bushel, what is the cost per ton of shelled corn (56 lbs. per bushel)? Of ear corn (72 lbs. per bushel)?
2. If it costs $16.20 per acre to grow a crop of corn and husk it

HORSES

from the standing stalks, what does it cost to produce a bushel of corn when it yields 40 bushels per acre? How much does such corn cost per ton of shelled corn? Per ton of ear corn?

3. If it costs $9.50 per acre to produce a crop of 2 tons of clover, how much does it cost per ton to produce clover hay?

FEEDING HORSES AT HEAVY LABOR

Requirements of a Horse.—A balanced ration for a horse at hard work must contain digestible nutrients in approximately the following amounts: .18 lbs. of protein, 1.2 lbs. of carbohydrates, and .05 lbs. of fat, per 100 lbs. live weight of horse. Thus a horse weighing 1,000 lbs. requires, when at hard labor, 1.80 lbs. of protein, 12.0 lbs. of carbohydrates and .5 lbs. of fat.

Combination of Feeds.—The proper combination of the feeds requires some thought and some figuring, but is a work that should be done on every farm or anywhere that horses are fed grain. Only by studying the feeding value of the different feeds, and their prices, can one be sure of the most economical ration. Sometimes it pays one to sell the feeds on hand and buy others in which the nutrients can be obtained more cheaply. As the market prices of the different feeds are changeable, one has a constantly varying problem. The wide awake feeder always has something to think about that is worth while.

Rations for a 1,200 lb. Horse at Heavy Work.—A 1,200 lb. horse at heavy work requires 2.16 lbs. of protein, 14.40 lbs. of carbohydrates, and .60 lbs. of fat.

The following rations will approximately supply these requirements: See table on page 165.

Ration No. I

	Pro.	C. H.	Fat
Oats, 16 lbs.	1.712	8.04	.608
Timothy hay, 14 lbs.	.392	6.08	.196
Total nutrients	2.104	14.12	.804

Ration No. II

	Pro.	C. H.	Fat
Corn, 15 lbs.	1.185	10.00	.64
Clover hay, 14 lbs.	.994	5.30	.256
Total nutrients	2.179	15.30	.896

Ration No. III

	Pro.	C. H.	Fat
Corn, 10 lbs.	.79	6.69	.43
Bran, 7 lbs.	.83	2.94	.17
Timothy hay, 14 lbs.	.392	6.08	.196
Total nutrients	2.012	15.71	.796

Each of the rations given above supplies approximately the needed amount of each of the digestible nutrients, also nearly the same amount of bulk. One may reasonably conclude that, of these rations, the one that can be most cheaply and conveniently supplied will be satisfactory. A little figuring will convince any one that there is a considerable saving by feeding ration No. II over feeding No. I, and that usually No. III will be cheaper than No. I.

If one is feeding a heavy ration, as one of the above, the grain should be reduced considerably, if a horse is to be idle for a few days.

A change of feed occasionally is probably better than to feed continually any one ration, as an animal appreciates a change

Figure 80.—Bundle corn is a cheap feed.

of diet fully as much as a person. An occasional feed of bran, when horses are fed timothy hay and oats, is a great benefit to a horse, as it aids in regulating his bowels. The addition of clover hay to the ration cheapens it, and adds greatly to its value, if the clover hay is bright and well cured. Some feeders are prejudiced against clover hay for horses, but good clover hay fed in moderation is a very desirable kind of feed for them.

Questions:
1. What do you understand by a balanced ration?
2. State the requirement, in digestible nutrients, of a horse at hard labor.
3. What can you say regarding the proper combination of feeds for horses?

Arithmetic:
1. If oats are worth 40c. per bushel and timothy hay is worth $6.00 per ton, what is the cost of Ration No. I?
2. If corn is worth 54c. per pushel (56 lbs.) and clover hay is worth $6.00 per ton, what is the cost of Ration No. II?
3. If bran is worth $24.00 per ton, corn 54c. per bushel, and timothy hay $6.00 per ton, what is the cost of Ration No. III?

FEEDING HORSES WHEN IDLE

Idle Horses.—Farm horses are idle or do very little work for a considerable portion of the year, and when they are idle one can greatly reduce the amount of grain they get and increase their allowance of hay. They do not need so much to eat, because they are expending very little energy. They have plenty of time to eat roughage, and, as roughage is cheaper than grain, it cheapens the ration very much to reduce the proportion of grain.

Figure 81.—A good team of farm mares at work. If much of the farm work is done with mares, and they are allowed to raise colts, the cost of horse labor can be materially cheapened.

Maintenance.—All that a mature horse needs when idle is enough to maintain his body. This is called a maintenance ration. If the horse is poor when the fall work is finished, he will need enough more than the maintenance ration to enable him to add to his weight the amount necessary to put him in good condition. A young horse, three to five years old, is still growing, and will need enough more than a maintenance ration to enable him to supply the needs of his body in growing.

There is always team work to do on the farm in the winter; and if several horses are kept some of them may be used for the winter's work and fed accordingly. The others should be kept over as cheaply as possible to maintain them in fair condition.

If kept comfortable, horses will do very well on just hay, preferably clover, and cornstalks. If it is desired to increase their weight, a little bundle corn may be fed in place of the stalks. These make one of the cheapest farm rations for wintering horses.

Ration of Bundle Corn and Clover.—Corn may be grown, cut shocked and hauled in from the field for about $14.75 per acre. If it yields 40 bushels per acre, there will be 2,240 lbs. of corn and probably about 3,000 lbs. of cornstalks. These cornstalks are not the best kind of feed, as they are too mature to be easily digested; but they add bulk to a ration. Maintenance requirements for horses have not been so carefully worked out as for cattle, and standards vary from .06 to .07 lbs. of protein and .6 to .7 lbs. of carbohydrates per 100 lbs. live weight of horse. A 1,200 lb. horse requires, when idle, about .8 lbs. of protein, 8 lbs. carbohydrates and .1 lbs. of fat. The following combinations of common farm feeds will be found to supply about the nutrients needed. See table on page 165.

Ration No. I

	Pro.	C. H.	Fat
Timothy hay, 15 lbs.	.42	6.51	.21
Oats, 4 lbs.	.428	2.00	.142
	.848	8.51	.352

Ration No. II

	Pro.	C. H.	Fat
Corn stover, 15 lbs.	.21	4.68	.105
Clover hay, 8 lbs.	.568	3.02	.144
	.778	7.60	.249

Ration No. III

	Pro.	C. H.	Fat
Bundle corn, 14 lbs.	.59	6.13	.317
Mixed hay, 8 lbs.	.396	3.25	.128
	.986	9.38	.445

Bundle corn is about 4-7 ear corn and 3-7 stover. Ear corn is 80% shelled corn and 20% cob. So 14 lbs. of bundle corn contains 6.4 lbs. of corn, 1.6 lbs. of cob and 6 lbs. of stover. Mixed hay is assumed to be half clover and half timothy.

Ration No. I represents a very common ration for idle horses. This ration would be changed very little, if good slough hay or upland hay were used in place of the timothy. Ration No. II is a little cheaper than No. I, and will give good results, if both the stover (cornstalks from which ears have been husked) and clover are good.

Ration No. III provides more nutrients than the others, and will supply the needs of a 1,200 lb. horse that is growing or gaining in weight, or maintain a horse that is doing some light work about the farm, as hauling straw or manure for a few hours occasionally. It is assumed that 14 lbs. of bundle corn contain 6 lbs. of corn and 8 lbs. of corn stover, and that the mixed hay is half clover and half timothy.

Questions:
1. Why do horses require less when idle than when at work?
2. What is meant by a maintenance ration?
3. What can you say regarding some economical combinations of feed for idle horses?

Arithmetic:
1. What is the cost of Ration No. 1, if timothy hay is worth $6.00 per ton and oats are worth 38c per bushel?
2. What is the cost of Ration No. 2, if corn stover is worth $3.00 per ton and clover hay is worth $6.00 per ton?
3. What is the cost of Ration No. 3, if bundle corn costs $4.00 per ton and mixed hay is worth $6.00 per ton?

CHAPTER XV
CATTLE
TYPES AND BREEDS

Breeds.—There are a great many different breeds of cattle, just as there are many different nationalities or races of people. Each breed seems well adapted to the particular community in which it was developed. Animals vary in character with the conditions under which they have developed, which accounts for the fact that all the breeds of cattle have been developed from the original wild cattle. Two men may start out with herds of cattle that are very similar. One will go to a cold, rugged climate, where feed is scarce and conditions are severe. The other may go to a warmer climate, where an abundance of feed is produced. Each man will have a different ideal and each will select the animals each year that will best please him. After several generations of cattle have been produced the two herds will be quite different.

Classes of Stock.—Every boy should know the common breeds of cattle, at least by sight and name and also know to what class each belongs. Cattle are commonly classed as beef cattle, those selected and raised only for meat, dairy cattle, those selected and raised only for milk and butter production, and dual purpose or general purpose cattle, those raised for both dairy products and beef. By far the greater number of cattle in the United States do not belong to any breed, because they have not been carefully bred for any particular purpose, and as a result are by no means uniform in character as pure-bred cattle are. Such cattle are called scrubs.

Pure-bred cattle have been raised for a long time, that is, for many generations, by men who have had a definite object or ideal in view. All the animals that have not appeared or performed up to the ideals of their owners have been sold. Only the select animals were kept. After following this practice for very many years all, or very nearly all, the animals within a breed are quite similar. These animals are then recognized as a distinct breed of cattle, and are called pure-bred. They are then given a name that indicates the country or community in which they were developed, or some characteristic of the breed.

Beef breeds are those bred particularly for meat. Beef cattle have short legs, deep bodies and broad backs. They have been selected for their meat-producing qualities and consequently have a tendency to lay on flesh, that is, fatten

Figure 82.—A shorthorn bull, a beef type.

more easily than animals selected for milk production. Beef cows usually give very much less milk than cows of the strictly dairy breeds. Shorthorns are the most common of the beef breeds. They originated in northeast England, in the counties of Durham and York. On this account they are often called Durham cattle. Shorthorns are red, red and white, white or roan in color, and are one of the large breeds. Some families of Shorthorns have been selected for milk production and are called "Milking Shorthorns." Herefords, another distinct beef breed, originated in Herefordshire, England, from which they get their name. They are also a heavy, large breed and are easily recognized by their color, red with white faces. They are sometimes called "Whitefaces." Aberdeen Angus cattle originated in Aberdeenshire, Scotland. They are polled (hornless) cattle, jet black in color. They are very round, black cattle, somewhat lighter in weight than Shorthorns, but are thick

Figure 83.—A group of Herefords, a beef type.

fleshed and are regarded as a very good beef breed. Galloways, another black polled breed of cattle, were originated in the hilly section of southwest Scotland. They have very short legs, heavy shaggy coats of hair, are very rugged and inclined to be wild. They are smaller than the Aberdeen Angus cattle, do not mature quite so young, but make a very good quality of beef.

Dairy Breeds have been selected especially for large milk production. Compared with beef cattle, they are more angular in form, with narrow backs, thin necks, large udders, and tend to produce milk rather than to lay on flesh.

These breeds are well adapted to farms where milk is the chief product desired. They are generally lighter in weight than the beef breeds. Holsteins are the largest of the dairy breeds, are black and white spotted in color and give the largest amount of milk per cow of any breed. Their milk tests low in butter-fat. They originated in Holland, and are gaining great popularity in this country. Guernseys

Figure 84.—A group of fine dairy cows.

are another favorite dairy breed. They are smaller than Holsteins and give richer, but less milk. They are yellowish or reddish fawn in color, often with some white spots, especially on the underline. They originated on the Island of Guernsey in the English channel. Jerseys also originated in the English channel, but on the Island of Jersey. They are smaller than Guernseys, are fawn colored, but may be either light or dark. They are more widely distributed in

Figure 85.—A Guernsey bull.

this country than any other dairy breed. They are noted for the richness of their milk. Ayreshires are a very beautiful breed of dairy cattle, originated in the county of Ayr in Scotland. They are red and white, or brown and white in color and are about the same size as the Guernseys, but are smoother in form and have more of a tendency to lay on flesh than any of the other dairy breeds. Their milk is

Figure 86.—A Guernsey cow, a dairy type.

richer than that of Holsteins, but not so rich as Jersey or Guernsey milk.

Dual purpose breeds are bred for both milk and beef; consequently they are neither strictly beef nor strictly dairy type but are part of each. Most of the scrub cattle in the country would be classed as dual purpose. Red Polled cattle are the only pure-bred cattle that are usually placed in this class. They originated in England, are medium in size, entirely red in color and have no horns. Many of the cows of this breed are excellent milkers, and, when fattened, animals of this breed produce very good beef. Shorthorns that have been selected particularly for milk are included in this class.

Other breeds of less importance in this country are the Dutch Belted, Brown Swiss, Devon, Kerry and French Canadian.

Figure 87.—A Holstein cow, a dairy type.

Questions:
1. What do you understand by the term pure-bred cattle? How do they differ from scrubs?
2. Name some of the characteristics of beef cattle, and describe four beef breeds.
3. Name some of the characteristics of dairy cattle, and describe four dairy breeds.

Arithmetic:
1. A keeps 5 common cows. Each gives 4,000 lbs. of 4% milk in a year. How much is the product worth at 32c. per pound of butter-fat? (4% milk is milk containing 4 lbs. of butter-fat per 100 lbs.)
2. B keeps 5 pure bred dairy cows. Each gives 6,000 lbs. of 5% milk in a year. How much is the product worth at 32c. per pound of butter-fat.

Figure 88.—A Jersey cow, a dairy type.

3. A sells ten beef steers, weighing 1,400 lbs each. How much do they bring at 7c. per pound?

4. B sells ten steers, weighing 1,100 lbs. each. How much do they bring at 5c per pound?

Figure 89.—A Shorthorn cow and calf, a dual-purpose animal.

CARE AND MANAGEMENT.

A farmer may have the best specimens of dairy cattle or beef-producing breeds, and yet, if he does not use judicious care in the management of them, the results are likely to end in loss and disappointment.

In addition to the general suggestions given in the chapter on live stock, the following points in regard to the care and management of cattle deserve attention.

Shade.—In summer it is very desirable that cattle be provided with the natural shade of trees, if possible. This promotes their comfort. Without the distraction of heat and flies dairy cows give more milk and beef cattle fatten more rapidly. If there are no trees in the pasture, artificial shade should in some way be furnished.

Water.—Both milch cows and steers should be well supplied with fresh clean water and never be allowed to drink from stagnant pools or mudholes. If running water is not convenient, well water is good, if supplied in plentiful quantities, preferably by a wind pump.

Our best dairymen find that it pays to take the chill off water for the cows. A very little fuel in a tank heater will take off the chill of the water. If it is not warmed in this way, the expensive feed that the cow eats will be used to warm the water. Besides, on a cold day a cow will not drink as much ice-cold water as her system requires.

Salt.—If salt is not kept constantly before cattle, it should be given to them regularly and frequently. It ought to be so placed as to be easily accessible and where it will be protected from rain.

Shelter.—It is especially true of milch cows and young stock that they require comfortable shelter. This is not simply a matter of comfort, but of dollars and cents. Steers do not require so warm quarters as long as they are protected from the storms and the wind; but they should be kept dry and given good bedding. Steers are fed more heavily and are fatter. They are thus better able to withstand cold.

Disease.—Occasionally, as recently, the "foot and mouth disease" makes its appearance with great disaster to cattle. No remedy is known for this epidemic except the slaughtering and burial of the affected animals and the maintenance of the most perfect sanitary conditions. The ordinary disease which is most prevalent among cattle is tuberculosis. This is a germ disease and is likely to originate in poorly ventilated and unclean buildings. Hence the necessity for light, ventilation and cleanliness. To guard against this disease it is best to kill all affected animals and to have each newly purchased animal tested before it is brought on the farm. There is danger in keeping one diseased animal that it will infect others through feeding troughs and grass.

Exercise.—All animals need some exercise, but milch cows should not be left outside to shiver in the cold. One practical dairyman says, "Leave the cows out no longer than you care to stand out in the same place with no overcoat on and nothing to do." If one follows this rule, cows will be left out but a short time in the cold or wet weather. Keep the cows in the barn most of the time during the winter, and give them a good bed and plenty to eat and drink.

Flies.—It is important to protect cattle as much as possible from flies. Relief is afforded by darkening the stable with curtains dropped over the windows. These may be made from burlap or old sacks. An effective spray, made of three parts of fish oil to one part of kerosene, may be applied as needed.

Bedding.—Plenty of good bedding is not only essential for the comfort of cattle, but it pays for itself in value as fertilizer. Whatever the bedding used may be, it should be kept under cover so that it is dry and sweet. In this condition it will keep the barn sweeter and freer from dust and will also absorb more liquid.

Questions:
1. Cattle naturally seek a shady place in hot weather. Why?
2. Why must cattle have plenty of clean water to do well?
3. Why do fattening steers not require so warm shelter as milch cows?
4. In what ways can we prevent disease in cattle?
5. What are the advantages of good bedding?

Arithmetic:
1. If six head of cattle worth $50 each were condemned and half the value paid by the state, what would be the farmer's loss?

CATTLE

2. If a cow drinks 90 pounds of water in a day, how many gallons would a herd of 12 cows drink in a day? (A gallon of water weighs about 8⅓ lbs.). How long would it take a boy to pump sufficient water for these cows if he could pump four gallons in a minute?

FEEDING

REQUIREMENTS OF DAIRY COWS

Maintenance.—Every animal requires a certain amount of feed for bodily maintenance, even though it may be idle, to keep up the body heat, for digestion, and for the other functions of the body. The requirements for maintenance are approximately the same for all animals of the same class and weight and kept under similar conditions.

The amount of such feed has been determined by feeding for several months, mature, idle animals kept under normal conditions, and by weighing the feed fed and weighing the animals, and by regulating the feed so that the various animals neither gain nor lose in weight.

Prof. T. L. Haecker, of the Minnesota Experiment Station, has found by extensive tests that .07 of a pound of protein, .7 of a pound of carbohydrates, and .01 of a pound of fat are required per hundred pounds live weight to maintain a cow not giving milk. If a cow weighs 1,000 pounds, it will require ten times these amounts for maintenance.

Nutrients Required.—Prof. Haecker has found also that the more milk a cow gives the more feed she needs, and the richer the milk the more feed required to produce it. For the convenience of feeders he has compiled a table from which the following is taken. Any one, knowing the weight of a cow and the amount and richness of her milk, can easily determine the amount of nutrients she needs for maintenance and to produce milk.

Nutrients Required to Produce One Pound of Milk of a Given Per Cent of Butter-fat

Per Cent Fat	Protein	Carbohydrates	Fat
3	.042	.19	.013
3.5	.045	.21	.015
*4	.048	.23	.016
4.5	.051	.25	.018
5	.054	.27	.019
5.5	.057	.29	.020
6	.060	.31	.022

From the last table it is an easy matter to determine the amount of nutrients required to produce a given number of pounds of milk of a given per cent of fat. For example, a cow that gives 15 lbs. of 4% milk will require, for its production, 15 times as much of the nutrients as is required to produce one pound of milk of the same quality. (See star in table on the preceding page.)

Fifteen times .048 lbs. of protein, .23 lbs. of carbohydrates and .016 lbs. of fat equals .72 lbs. of protein, 3.45 lbs. of carbohydrates and .24 lbs. of fat. Thus, it will be seen that a cow requires .72 lbs. of protein, 3.45 lbs. of carbohydrates and .24 lbs. of fat, simply for the production of milk. In addition to this, she must be supplied with feed for bodily maintenance. If the cow weighs 1,100 lbs. she will require 11 times .07 lbs. of protein, .7 lbs. of carbohydrates and .01 lbs. of fat (the amount required to maintain 100 lbs. live weight) or .77 lbs. of protein, 7.7 lbs. of carbohydrates and .11 lbs. of fat. We know, then, that a cow weighing eleven hundred pounds, and giving 15 pounds of 4% milk requires daily:

	Pro.	C. H.	Fat
For maintenance	.77	7.7	.11
For 15 lbs. of 4% milk	.72	3.45	.24
Total daily requirement	1.49	11.15	.35

From the above it will be seen that, to determine the requirements of a cow, one must know approximately her weight, her daily milk production and its per cent of fat. If these facts are known, it is, with the table, a very simple mathematical problem to determine her daily needs.

Questions:
1. What do you understand by the term "food of maintenance?"
2. How have feeders found out how much feed animals require for maintenance?
3. Why does a cow require more feed when giving 20 lbs. of 4% milk than when she is giving 10 lbs. of 4% milk?
4. What three factors must be known in order to determine the daily feed requirements of a cow?

Arithmetic:
1. For bodily maintenance a cow requires .07 lbs. of protein, .7 lbs. of carbohydrates and .01 lbs. of fat per 100 lbs. live weight. How many pounds of each nutrient are required to maintain a cow weighing 1,050 lbs?

CATTLE 193

2. If .051 lbs. of protein, .25 lbs. of carbohydrates and .018 lbs. of fat are required to produce 1 lb. of 4½% milk, how many pounds of each nutrient are required to produce 18 lbs. of 4½% milk?

3. How much protein, carbohydrates and fat will a 1,050-pound cow giving 18 lbs. of 4½% milk require?

TO COMPOUND A RATION

Proportion of Grain to Roughage.—When one knows a cow's requirements it is a very simple matter, by using the table on page 165, showing the composition of feeds, to compound a ration that will supply them.

We have found that an eleven hundred pound cow giving 15 lbs. of 4% milk daily requires daily 1.49 lbs. of protein, 11.15 lbs. of carbohydrates and .35 lbs. of fat. A cow could not eat enough roughage to supply this amount of nutrient. She must have some more concentrated feed such as grain. Many dairymen feed grain in the proportion of 1 lb. of grain to each 3 lbs. of milk that the cow gives, and supply the rest of the nutrients required by feeding roughage. This is practically a safe basis. Thus the cow, whose record is given above, would require about 5 lbs. of grain (as she gives 15 lbs. of milk daily) and roughage to complete the ration.

A Simple Ration.—To compound a ration one must know the composition of various feeds to be fed. See table, page 165.

Daily Ration for 1,100 Pound Cow Giving 15 Lbs. of 4% Milk Daily

	Pro.	C. H.	Fat
Corn, 4 lbs.	.316	2.67	.172
Bran, 1 lb.	.119	.42	.025
Clover hay, 10 lbs.	.710	3.78	.180
Fodder corn, 10 lbs.	.37	4.14	.146
Total nutrients	1.515	11.01	.523

It will be seen that this ration contains approximately the right amount of protein, for which no other nutrient may be substituted, but is a little deficient in carbohydrates. There is .17 lbs. more fat than is required, which may be used to make up the shortage in carbohydrates. We have learned that fat and carbohydrates are used for the same purposes in the animal body, and that fat is worth 2.2 times as much as carbohydrates, hence the excess fat (.17x2.2) is equal to .37 of a pound of carbohydrates; which, added

13—

to the 11.01 lbs. furnished by the ration, makes 11.38 lbs. or approximately what is required.

Features of the Previous Ration.—It is not usually desirable to feed as large a proportion of corn to dairy cows as is provided by this ration, as corn has a tendency to produce fat rather than milk. But when clover hay, which is rich in protein, is fed, a large proportion of corn or other non-nitrogenous grains may be used. When timothy,

Figure 90.—Mixing the grain feed consisting of 100 lbs. of bran and 400 lbs. of cornmeal.

corn stover or wild hay, which are deficient in protein, are fed for roughage, a larger amount of bran or other nitrogenous grain would be required.

Since clover can be grown as cheaply as any hay crop, it is advisable to provide plenty of well cured clover hay; for it makes possible the use of cheaper grain feeds, as corn and barley, instead of oats, bran and oil meal, which must be fed, if non-nitrogenous roughage is used.

Feeding a Ration.—To feed a cow such a ration, it is not necessary to weigh each day 4 lbs. of corn, 1 lb. of bran and the hay and fodder. One would mix 100 lbs. or more of bran with four times as much corn. In feeding, use a measure that holds the desired number of pounds of the mixture. By weighing the feeds a few times one can soon approximate the right amount of each without weighing it.

Questions:
1. In what way can one determine approximately the amount of grain needed by a cow?
2. Tell how to compound a ration for a cow.
3. Would you weigh the ration each time you feed?

Arithmetic:
1. What is the cost of a ration composed of 4 lbs. of corn at 54c. per bushel, (56 lbs.), 1 lb. of bran at $20 per ton, 10 lbs. of clover hay at $5.00 per ton, and 10 lbs. of corn fodder at $4.00 per ton?
2. A cow fed the above ration gives 15 lbs. of 4% milk. What is the milk worth when butter-fat is selling for 30c. per pound?
3. How many pounds of protein in one ton of bran? What does it cost per pound when bran costs $20 per ton?
4. How many pounds of protein in one ton of clover hay? What does it cost per pound when clover hay costs $5.00 per ton?

A Poor Ration.—A very common ration fed to dairy cows on the farm is composed of slough hay, corn stover, and ground barley and corn. Suppose a 1,000-lb. cow gives 20 lbs. of milk testing 4.5% fat. She will require:

	Pro.	C. H.	Fat
For maintenance	.7	7.	.1
For 20 lbs. of 4.5% milk	1.02	5.	.36
Total requirements	1.72	12.	.46

A cow giving 20 lbs. of milk would require 7 or 8 lbs. of grain. Let us see what kind of a ration she would get, if fed the above mentioned feeds:

Daily Ration for a 1,000 lb. Cow Giving 20 lbs. of 4.5% Milk

		Pro.	C. H.	Fat
Corn	4 lbs.	.316	2.67	.172
Barley	3 lbs.	.252	1.96	.048
Slough hay	12 lbs.	.312	5.03	.132
Corn stover	12 lbs.	.168	3.74	.084
Total nutrients		1.048	13.40	.536

This ration would be all a cow could possibly eat, as it supplies 31 lbs. of dry feed, yet it comes far short of supplying enough protein. We have learned that neither carbohydrates nor fats can take the place of protein. A cow must have sufficient protein or she cannot maintain her body or produce milk. It is not to be wondered at, then, that a cow fed a ration similar to the above would gradually decrease in her milk flow until her requirements balanced the protein she was getting. She would then get more carbohydrates (fat-forming feed) than needed for milk production, and so would begin to fatten, an undesirable result in a milch cow.

A Good Ration.—The required nutrients for the cow mentioned above could be supplied by replacing 3 lbs. of the corn or barley with 3 lbs. of oil meal (try it); but such a ration would be more expensive and no better than if the 12 lbs. of slough hay and 7 lbs. of the corn stover were replaced with 16 lbs. of clover hay. This would make a good ration and supply the need of the cow.

Daily Ration for a 1,000 lb. Cow Giving 20 lbs. of 4.5% Milk

		Pro.	C. H.	Fat
Corn	4 lbs.	.316	2.67	.172
Barley	3 lbs.	.252	1.96	.048
Corn stover	5 lbs.	.070	1.56	.035
Clover hay	16 lbs.	1.136	6.04	.288
Total nutrients		1.774	12.23	.533

This ration emphasizes the fact that every farmer should provide plenty of good clover hay for his cattle.

Questions:
1. Can you tell what is wrong with the ration given under the head "A poor ration?"
2. Why would a cow be unable to keep up her flow of milk, if fed such a ration?
3. Why is the ration given under the head "A good ration" better than the other?

Arithmetic:
1. Find the cost of the grain in the above ration, 4 lbs. of corn and 3 lbs. of barley. (Corn 50c. per bushel (56 lbs.), and barley 50c. per bushel (48 lbs.).)
2. Find the cost of a similar ration, if 3 lbs. of corn were replaced with 3 lbs. of oil meal, making the grain ration 1 lb. of corn, 3 lbs. of oil meal and 3 lbs. of barley. (Oil meal costs $35.00 per ton.)
3. What is the entire cost of the last ration given, corn 50c. per bushel, barley 50c. per bushel, clover hay $5 per ton and corn stover $3 per ton?
4. Find the requirements of a 1,200 lb. cow, giving 35 lbs. of milk testing 3.5% fat.

CATTLE

SUCCULENT FEED

Kind of Ration.—On page 192, a ration was suggested for an eleven hundred pound dairy cow giving 15 lbs. of 4% milk. This ration consisted of 4 lbs. of cornmeal, 1 lb. of bran, 12 lbs. of clover hay and 10 lbs. of fodder corn. It supplied all the nutrients in the proportion needed by the cow, and, if the cow were made comfortable and regularly fed and milked, she would do fairly well. Such a ration is more economical and will give better returns than the ordinary ration of timothy or wild hay, corn stover and some of the common farm grains, as barley or corn.

Figure 91.—Filling a silo at University Farm. Corn is cut while still green, when ears are well glazed, hauled directly, cut and put in the silo. Silage does not spoil, because air is kept away from it. It is still green and succulent when fed.

Cows Need Succulent Feed.—We know that cows usually give the most milk when they are in good pasture. Their chief feed is then green grass. This fact would indicate that such feed is better for milk production than the dry feeds fed in winter. Green feed is more easily digested than dry, coarse fodder, such as hay, fodder corn and corn stover. Less energy, moreover, is required to digest it, it tends to keep the body and digestion in better condition, and it stimulates the appetite. In the winter, when fresh vegetables are scarce and we eat potatoes, bread and meat for a long time, we become tired of them and crave something succulent, as fruit and green vegetables. In well regulated homes such food is supplied by canned or fresh vegetables and fruits. The barrel of apples in the cellar is not exceedingly valuable from the standpoint of amount of nourishment it contains but from the succulence and refreshing effect of the apples. In like manner it pays to supply the live stock on the farm with something to take the place of the green grass they get in the summer. The whole ration need not be of succulent material, but that a

Figure 92.—A load of mangels. Mangels may be grown at from $1.60 to $2.00 per ton. They are a very valuable addition to the dry feeds commonly fed in winter, as they are palatable and succulent.

portion of it should be is quite essential to best results. Just as an apple each day is good for a boy or girl, so are a few pounds of succulent feed (such as roots or silage) each day good for farm animals.

Ensilage is one of the cheapest succulent feeds during the winter, that is, when one has a large herd and is able to build a silo and buy the necessary machinery for handling the crop. It is good feed, handy and very much relished by all classes of stock.

Ensilage is usually corn, (sometimes other crops), stored green in a large tank called a silo. The silo must be airtight or nearly so, as the green feed is kept from spoiling by keeping the air away from it—just as berries are preserved in fruit jars.

Roots.—Another way by which succulent feed may be supplied is by raising roots, as mangels, rutabagas, stock carrots, etc. For small herds, roots are cheaper than ensilage, as no expensive machinery or storage room is required. By planting roots on rich land, fifteen to twenty tons may be grown per acre. One half an acre to two acres of roots well cared for will supply from eight to twelve cows.

Questions:
1. Why do cows usually give more milk in summer than in winter?
2. What is the difference in their feed in summer and winter?
3. What do animals need in winter in addition to dry feed?
4. How may succulent feed be supplied to animals in winter?
5. To what conditions is ensilage better adapted than mangels?
6. What do people eat in winter to supply succulent food?

Arithmetic:
1. If corn contains 89% dry matter and mangels contain 9% dry matter, how many pounds of mangels are required to supply as much dry matter as is supplied by 5 lbs. of corn?
2. If mangels contain 9% dry matter, how many pounds of water in 100 lbs. of mangels?
3. If mangels yield 20 tons per acre, how many tons of dry matter are produced per acre? (Mangels are 9% dry matter.)

RATIONS CONTAINING SUCCULENT FEED

Composition of Feeds.—To intelligently compound rations with ensilage or roots forming a part, it is necessary to know the amount of digestible nutrients in the various feeds used. The following table gives their composition in the two rations to be compounded.

A glance at the table will show that a pound of corn silage contains about one third as much digestible nutrients as fodder corn.

This proportion is due to the fact that the ensilage has practically as much water in it as when it was cut green in the field, while the fodder corn has been dried out (cured). In other words three pounds of green corn, cut and shocked in the field, will dry out and make about one pound of fodder corn. We would need, then, to feed about three times as many pounds of ensilage as of fodder corn to get the same amount of nutrients. But ensilage has the additional value of stimulating digestion and keeping the animal in a good healthy condition.

Ration Containing Ensilage.—The ration suggested on page 192, for an 1,100-pound cow giving 15 pounds of 4 per cent milk, as suggested, consisted of 4 pounds of cornmeal, 1 pound of bran, 12 pounds of clover hay and 10 pounds of fodder corn. This supplied approximately the nutrients needed by the cow, which are 1.49 pounds of protein, 11.15 pounds of carbohydrates and .35 of a pound of fat.

Ration Containing Ensilage, for 1,100 Pound Cow Giving 15 Pounds of 4 per Cent Milk

		Pro.	C. H.	Fat
Cornmeal	4 lbs.	.316	2.67	.172
Bran	1 lb.	.119	.42	.025
Clover hay	10 lbs.	.710	3.78	.180
Corn silage	30 lbs.	.37	4.30	.210
Total nutrients		1.515	11.17	.587

The above ration furnishes approximately the nutrients required. It differs from the ration given on page 192 only in containing thirty pounds of corn silage in place of ten pounds of fodder corn. So far as nutrients are concerned, there is very little difference; but a cow would give better returns on this ration than on the former, owing to the succulence added by the ensilage, which makes the whole ration more palatable and more easily digested.

Ration Containing Roots for 1,100 Pound Cow Giving 15 Pounds of 4 per Cent Milk

		Pro.	C. H.	Fat
Corn	4 lbs.	.316	2.67	.172
Bran	1 lb.	.119	.42	.025
Clover hay	10 lbs.	.710	3.78	.180
Mangels	20 lbs.	.200	1.10	.040
Fodder corn	8 lbs.	.296	3.31	.117
Total nutrients		1.641	11.28	.534

It will be seen that mangels do not provide so much nutrient matter per pound as the ensilage, but are a little more valuable as a means of furnishing succulence, as they are sweet, while ensilage is more or less sour.

CATTLE

To get enough carbohydrates in this ration it was necessary to add some fodder corn. A ration containing both clover hay and fodder corn can usually be balanced by changing the proportion of these two feeds. Fodder corn is rich in carbohydrates and clover hay in protein.

Questions:

1. How do fodder corn and corn silage compare in feeding value?
2. For what reason is a ration containing ensilage better than one containing only cured roughage?
3. In what respect do mangels and ensilage differ?

Arithmetic:

1. Find the daily requirements of protein, carbohydrates and fat for a 1,000-lb. cow giving 20 lbs. of 4% milk. See page 192.
2. Find the amount of protein, carbohydrates and fat in 5 lbs. of cornmeal, 2 lbs. of bran, 12 lbs. of clover hay and 9 lbs. of fodder corn. (See page 165.)
3. Find the amount of protein, carbohydrates and fat in a ration the same as the above, but replace the 9 lbs. of fodder corn with 27 lbs. of corn ensilage.
4. Find the amount of fodder corn and mangels required to furnish about the same amount of nutrient as 27 lbs. of silage.

CHAPTER XVI
DAIRYING

MILK AND ITS CARE

Milk.—Milk is nature's perfect food for animals. It consists of water 87%, sugar (carbohydrates) 5%, fat 4%, casein and albumin (protein) 3.3%, and ash .7%. In 1909 nearly seven and a half billion gallons of milk were produced on the farms of the United States, the value of the product of which, excluding home consumption, was almost six hundred millions of dollars. The average number of gallons produced per cow was 362. It is readily seen, therefore, that in comparison with the record given below, there is much room for improvement in the amount that might be realized. At present the world's championship cow, Finderne Pride Johanna Rue has produced 28,403.7 pounds (3,664.3 gallons) of milk in one year, an average of over 10 gallons a day. Of this amount 1,176.47 pounds was butter-fat. This cow is a Holstein.

Milk holds the tiny particles of butter-fat in suspension. Such a liquid is called an emulsion. The butter-fat is lighter than the rest of the milk, and, therefore, rises to the surface; but milk as a whole, is .032 heavier than water.

As human food milk contains all the elements necessary to the human body and in very nearly the proportions needed. It contains protein, muscle-forming material, in casein; carbohydrates, heat and energy-forming material, in fat and sugar; and ash, or mineral matter, needed in bone-building. The average American family spends from two to four times as much for milk, cream and butter as for bread. Cow's milk is a very good substitute for mother's milk for babies. In fact pure, fresh milk is almost necessary for growing children.

On account of the importance of milk and milk products as human food it is very important that they be handled in a most careful manner.

Care of Milk.—As milk is so extensively used as a human food, the proper care of it needs to be emphasized. Attention should be given to

The Stable.—This must be well lighted and ventilated to assist in keeping it free from bacteria and odors. Dust also should be prevented as much as possible, and coverings should be provided for the pails while in the stable.

The Cow.—She should be kept scrupulously clean, and her udder and teats should be washed before milking. All milch cows should be curried.

The Milker.—The person who milks should wear clean clothes and have clean dry hands. The hands should never be put in the milk.

The Utensils.—Porcelain or smooth tin pails without open seams are best. These should be scalded after every milking and allowed to stand in the sun and air. The churn should be scalded also, as well as the parts of a separator, and all bottles.

Keeping.—Milk should be removed from the stable as soon as possible and chilled. Low temperatures are unfavorable to bacteria. The milk should be kept where it will not be possible for it to absorb any unpleasant odors, as it is inclined to do.

Souring.—Very minute plants, called bacteria, find in milk a fertile field for development. In their growth they change the milk sugar to lactic acid, which "sours" the milk and curdles it. This may be prevented by extreme cleanliness and low temperature which does not favor the growth of bacteria.

Pasteurization.—Milk may be kept longer by heating it for twenty minutes at a temperature of 160° F. and then cooling it rapidly. This process is called Pasteurization. It destroys any bacteria that may be present. In many cities there is a law that milk must be so treated. When milk is heated above 180° F., as at that temperature there is extra certainty that the bacteria will all be killed, it is said to be "sterilized."

Sanitary plants, to produce absolutely pure milk, take all these precautions.

Questions:

1. A pound of milk is said to possess practically as much nourishment as a pound of beefsteak. Which is cheaper?
2. How many pounds or gallons of milk are your cows or others in your community giving per day? Per year? Is it more or less than the average?
3. When all things are considered, does it cost more to keep a good cow than a poor one?
4. Why is it that milk is so perfect a food?
5. Why is it important to be so particular in the care of milk?

Arithmetic:

1. If butter-fat is worth 28c. per pound, what was the value of the butter-fat produced in one year by the World's Champion Cow?
2. If the cow that gives 362 gallons of milk a year produced 4.14% of butter-fat, what is the value of her butter-fat compared with the champion?
3. If patrons will pay 1c. a quart extra for good clean milk, how much extra might be received from the champion's 3,644 gallons?

TESTING MILK

Babcock Test.—The Babcock Test is one of the great inventions of the age, and has done a great deal for the dairy industry. Before this invention there was no way to tell the different grades of milk. It was known that some cows gave richer milk than others, but the only way to tell the amount of butter-fat in a given sample was to raise the cream and make it into butter. When all the butter was made on the home farms, it was not so necessary to know the quality of milk; but when creameries became common, and milk was sold, it became important to know how much butter-fat each sample contained, so that it could be paid for in proportion to its value.

Dr. Babcock realized this need, and, after years of effort, invented a test by which any sample of milk may be tested in a very short time and its per cent of butter-fat determined. Thus it is possible for every farmer, who hauls milk to a creamery, to be paid for it in exact proportion to the amount of butter-fat it contains.

Principle of the Test.—This test is very simple, and makes use of a few facts that everyone knew before the test was invented. It was known that, if milk were set away for several hours, cream would rise to the top. This fact indicated that cream is lighter than the other parts of the milk. Every boy, who has ever turned a grindstone, knows that, if water is poured on a grindstone, and the stone turned rapidly, the water is thrown off. This indicates that anything revolving has a tendency to be forced away from the point around which it is revolving.

Dr. Babcock made use of these two principles by devising a machine in which bottles can be set and revolved rapidly. To make the test, a certain amount of milk is put in a test bottle and some sulphuric acid added. The acid assists in breaking down the milk, and makes it easier for the fat to be separated from it. The test bottles are put in the machine, which is then turned at a given speed. The cups in which the bottles are set swing outward, as the speed increases, until the bottles are in a horizontal position, with the bottoms the farthest away from the center, around which they are revolving. This rapid revolving tends to force all the milk into the bottom of the bottle, just as turning the grindstone tends to throw water away from it. Milk, being heavier than cream, is crowded with more force into the bottom of the bottles, and the fat rises into the necks. The acid added to the milk causes it to turn dark, and the butter-fat is amber colored, so that the fat is easily distinquished from the milk. The necks of the bottles have scales on them, and as the cream is forced into them, one notices how many spaces are filled with fat. The number filled indicates the per cent of the fat in the milk. If three spaces are filled, the milk tests three per cent, and, if four and a half spaces are filled, the milk tests four and a half per cent.

Milk Test.—If we say milk tests five per cent fat, we mean that in one hundred pounds of milk there are five pounds of butter-fat. One hundred pounds of six per cent milk is worth twice as much at the creamery as one hundred pounds of three per cent milk.

At every creamery this test is used, and when a farmer brings milk to the creamery, the butter maker weighs it, takes a sample and tests it. The total weight multiplied by the test of the milk and divided by one hundred, gives the pounds of fat and afford the basis on which the milk is paid for. Thus, if a farmer delivers one hundred and seventy-five pounds of milk, and it tests four per cent, the problem is solved as follows: .04 x 175 lbs. = 7, the number of pounds of fat.

Cream is not pure butter-fat; so it must be tested also.

Questions:
1. Why was some device for testing milk badly needed?
2. Who invented such a device?
3. Upon what principles does it work?
4. If you have seen testing done at the creamery, describe it as best you can.
5. What is meant by per cent of fat in milk?

Arithmetic:
1. A hauls 230 lbs. of milk to the creamery. It tests 3.5% fat. How many pounds of butter-fat does it contain? How much is the butter-fat worth at 34c. per lb.
2. B hauls 150 lbs. of milk to the creamery. It tests 4.5% fat. How many pounds of fat does it contain? How much is the fat worth at 34c. per lb.?
3. C delivers 75 lbs. of cream to the creamery. It tests 24% fat. How many pounds fat has he? What is it worth at 34c. per lb.?

TESTING COWS

Culling.—Now that dairying is getting to be such an important part of farming, farmers are studying how they may secure greater profits. They have found that, if they are to realize satisfactory returns on dairying, they must keep only cows that are capable of producing large amounts of milk and butter-fat. They have found also that the only way to know just what each cow is doing is to weigh the milk. At first this work seemed unnecessary, but now nearly every good dairyman tests his cows and finds that it pays him to do so, because it takes less time than to milk one or two unprofitable ones.

Weighing Milk.—This is very simple, if one has a spring balance hung in the barn near the milk can, and a sheet of paper with a column for each cow, tacked up nearby so that the results can be jotted down as the milk is weighed. It is not necessary to weigh the milk every day. Weighing it night and morning, once every ten days, or even once a month, makes it possible to determine quite accurately the amount of milk a cow has given for the month.

Advantages of Weighing Milk Every Day.—There are a great many advantages, though, in weighing milk every day. By so doing one knows accurately how much milk a cow has given during the month. It also enables one to tell at once whether a cow gives less than her usual flow of milk. If a cow has been giving fifteen pounds of milk, and suddenly drops to thirteen pounds, one's attention is called to the fact at once and the cause can be sought. It may be found that the cow got out of the yard and that the dog was set on her; that she was left out in the cold too long; that she was turned out to drink when the wind was so cold and the water so nearly frozen, that she did not get the amount of water she needed, or that a window or door was left open and the cold draught chilled her. Whatever the cause of the loss in milk may be, if attention is called to it, the cause and the remedy may both be found.

Figure 93.—Scales and convenient case for weighing and sampling milk.

Sampling.—A sample to be tested for the per cent of fat should be taken just as the milk is being weighed. Stir the milk in the pail to make sure that it is all uniform; then take a small sample in a bottle. Each sample may be tested soon after being taken, or a simpler way is to take several samples from the same cow and keep them in one bottle, and test all together. In this way of testing, some preservative must be added to keep the samples from spoiling.

Testing.—Any careful boy, twelve or more years old, can test milk, if he has a tester and is shown how, or, if the milk is weighed at home and samples taken, the creamery man will, as a rule, test them for a very small charge or for nothing. At one place the farmers pay three cents a sample for testing. If one has a sample tested each month from a cow, it would cost but thirty-six cents per year, and this would certainly be a paying investment. In some places the farmers form a cow-testing association and hire a man to test all the cows. He comes to each farm once a month,

Figure 94.—A pure-bred Holstein cow. She gave in 30 days 114.9 pounds of butter.

and weighs and tests the milk from each cow night and morning, then goes to the next farmer. This is a very satisfactory arrangement.

Work for Boys.—Any boy who is milking cows can get good lessons in arithmetic, and greatly increase his knowledge of dairying, by doing the work here suggested. He can weigh the milk separately from each cow, take samples each month, and test them at home or have them tested at the creamery. He will ascertain how much milk each cow gives during the month, and the per cent of butter-fat. He can then (probably at school) figure out how much butter-fat each cow has given during the month. It will be valuable to compare the records of different cows, and to find out why cows are doing better on one farm than on another.

Figure 94a.—Outfit for making the Babcock test. From left to right, an eight bottle tester; a graduate for measuring acid; a test bottle; compasses for measuring fat in neck of bottle; a pipette for measuring milk, and a sample bottle.

Questions:

1. Why is it wise to weigh and test the milk from each cow?
2. What are the advantages in weighing the milk from each cow at each milking?
3. Tell how you would take a sample of milk for testing?

Arithmetic:

1. A cow gives an average of 20 lbs. of milk per day for 300 days each year. What is her yearly milk production?
2. A cow gives 6,000 lbs. of milk in a year, testing 4% fat. How many pounds of butter-fat foes she give? How much is the butter-fat worth at 30c. per pound?
3. A cow gives 6,000 lbs. of milk in a year, testing 5% fat. How many pounds of butter-fat does she give? How much is the butter-fat worth at 30c. per pound?

CHAPTER XVII

SHEEP

TYPES AND BREEDS

Types.—Sheep are kept for mutton and wool production. Some breeds are specially adapted to the production of wool and do not produce the best quality of mutton. Other breeds are well adapted to the production of mutton, but are not so well adapted to wool production. There are two common ways of classifying sheep, one, on the basis of wool, as fine wool, medium wool and long wool. The other, as fine wool and mutton. By the last classification the mutton breeds include the medium wool and long wool breeds.

Figure 95.—A Rambouillet, a fine wool type.

Fine wool breeds have been bred chiefly for wool production. The wool on these breeds is comparatively short, thick and very fine, and very oily, so that a great deal of dirt sticks to it. Their bodies are thin and irregular and their skins are very much wrinkled. These breeds are smaller generally than the mutton breeds. They have white faces and the males have horns and the ewes are hornless. American and Delaine Merinos are the most important breeds in this class. They were brought to America from Spain more than a hundred years ago. They have been generally recognized as American breeds. The American Merino is very much wrinkled. The Delaine Merino is wrinkled, but not so much as the former. Rambouillet, the other fine wool breed, originated in France. It is larger than the Merino, has a little better mutton carcass, is generally wrinkled very little and only at the neck.

Medium wool breeds have been bred chiefly for mutton, hence have very compact bodies, well covered with flesh. They produce good fleeces, but not as heavy or as fine as the fine wool breeds. The breeds of sheep in this class are

Figure 96.—A Shropshire, a medium wool type.

by far the most common in America. Five of the important breeds in this class are Shropshire, Southdown, Oxford, Hampshire, and Suffolk. These are all English breeds, with good carcasses. They always have brown or black faces and legs and no horns. Other breeds in this class are Cheviot and Dorset. These breeds have white faces, the Dorset has horns—both males and females. The Dorset is an English breed. The Cheviot comes from Scotland.

Coarse wool breeds have also been bred chiefly for mutton. They are large breeds of sheep, taller than the Shropshire, and with much longer, coarser wool. These breeds have good mutton carcasses and no horns. The Leicester, Cotswold and Lincoln are English breeds with white faces. The Blackfaced Highland is a Scotch breed.

Figure 97.—A Cotswold, a coarse wool type.

Questions:

1. Tell how breeds of sheep are classified as to purposes for which they are kept? As to the kind of wool?
2. What breed of sheep is most common in your neighborhood? Describe this breed and show how it differs from other breeds.
3. Name the breeds of sheep that have horns? In which breed do both males and females have horns?

Arithmetic:

1. What is the value of a fleece of wool weighing 7½ lbs., when wool is 23c. per pound?
2. What is the value of 17 lambs weighing 84 lbs. each, at 7c. per pound?
3. If a farmer with 50 ewes can get 7½ lbs. of wool from each, and can raise 40 84-lb. lambs, what will be his total income from his sheep, if he sells the wool at 23c. per lb. and the lambs at 7c. per lb.?

CARE AND MANAGEMENT

Care and management have quite as much to do with successful sheep husbandry as with the successful management of any kind of live stock. While sheep may be able to live with less attention and shelter than are required by other classes of live stock, they will not prove profitable unless made comfortable and given attention when needed.

During the winter the flock on the average farm consists chiefly of ewes kept over winter with the expectation that they will have lambs in the spring.

Winter Care of Breeding Ewes.—Ewes kept on good pasture during the summer and fall, are usually in good flesh by the time they are put in winter quarters. During the winter they need food enough to maintain their bodies, to provide for the growth of wool and to supply the energy needed for what little exercise they take about the sheds and yards. They do not need to be fed as heavily in proportion to their weight as cows that are giving milk, horses that are working, or cattle, sheep or hogs that are being fattened.

It is unwise to feed ewes any ration that will tend to fatten them. They should be fed succulent and muscle-forming foods, as clover hay, corn fodder, roots; and, if fed any grain, it should be a kind rich in protein, as bran, oats, etc.

To feed sheep properly is as much of a problem as to feed other kinds of stock. If sheep are not well nourished, they will lose some of their wool, and be weak and poor in the spring. If breeding ewes are fed too much, the lambs are likely to come weak in the spring; besides, there is an unnecessary waste of feed.

Require Little Labor.—Most of our farms need more stock than is at present kept on them. Farm labor is so scarce that it seems unwise, on the average farm, to increase the number of cows kept beyond what can be cared for, if necessary by the family. Sheep require comparatively little labor, except for a short time during the lambing season. Five to seven ewes will bring in about as much income in a year as a cow, and less labor is required to care for them. In view of these facts, the live stock of the farm may often be more easily and more profitably increased by putting on a flock of from twenty to seventy ewes than by adding from three to ten cows to the herd.

Shelter.—Sheep are so well protected by their wool that they need very little or no protection from the cold. They should, however, be kept dry and have a place, that is free from draughts, in which to lie down. A single board or

Figure 98.—Shearing sheep by machinery.

straw shed, closed tight on three sides, but with the other side partly open so that the sheep may run out and in at will, is a very satisfactory place in which to keep sheep. If lambs come during the cold weather, warm quarters must be provided.

Fencing for Sheep.—One of the objectionable features of keeping sheep on the average farm is the difficulty of fencing them in. A fence that can be built for 20 cents to 25 cents per rod is satisfactory for cattle. A much closer fence, as a woven wire fence or a narrow ribbon of woven wire with one or more barbed wires above, is needed for sheep. Such a fence costs 45 to 60 cents per rod. As farms are more intensively worked, however, more and better fences will be used, and then there will be no difficulty in keeping sheep.

Sheep for Fattening.—A number of farmers, who have not the necessary fencing so they can raise sheep to advantage, have gotten some of the benefits of having sheep on their farms by buying at about harvest time a carload of lambs, or as many as they can keep to advantage, and allowing them to graze over their stubble and cornfields during the fall. Such sheep are in fine condition to fatten during the winter on bundle corn or other cheap feed. Farmers may thus produce several pounds of mutton on each acre of land, after it has produced a crop; make their land cleaner and richer; and feed on the farm, at a profit, products otherwise of little or no value.

Questions:
1. What winter care do breeding ewes require?
2. Compare the shelter needed for sheep with the shelter needed for dairy cows.
3. What can you say about fencing for sheep?
4. Why do sheep require less labor than other stock?

Arithmetic:
1. A farmer buys 50 sheep at $4.50 each. How much do they cost him?
2. When shorn, the 50 sheep average $7\frac{1}{2}$ lbs. of wool each. How many pounds of wool will the farmer have? How much is it worth at 24c. per pound? How much is the wool worth per sheep?
3. From the 50 sheep the farmer raises 45 lambs worth $4.00 each. How much are the lambs worth? What is the average income for lambs from each of the 50 sheep?

FEEDING

Cheaply Raised.—Sheep can eat and thrive on a diet constituted principally of roughage. In this respect they are like cattle; but very different from hogs whose ration must be largely grain, because they have comparatively small stomachs. It is well to feed at least some of the grain produced on a farm for the sake of preserving its fertility; but there is always more or less roughage, as straw, cornstalks, hay, weeds and scattered grain. This coarse stuff is not suitable for dairy cows, although beef cattle can use such feed; but sheep are more likely to return a profit on such feed than beef cattle.

216 ELEMENTS OF FARM PRACTICE

Sheep produce two crops a year, a crop of wool and a crop of lambs. Often the wool will pay for a year's feed. The lambs will then be clear, and a lamb is ready for market in from six months to a year, while a calf is not profitable for two or three years.

Rations for Breeding Ewes.—Some sheep feeding work done by the Animal Husbandry Division of the Minnesota

Figure 99.—A good flock of breeding ewes gleaning in a stubble field. They will pick up all heads of scattered grain, as well as destroy many troublesome weeds.

Experiment Station, shows that with the common and cheap farm feeds, breeding ewes may be wintered with excellent results and very cheaply.

The rations given below show the amount of the different feeds per day per 100 lbs. live weight of sheep.

Ration No. I
3.7 lbs. of fodder corn, in which there were a few nubbins of corn.

Ration No. II
3.7 lbs. of second crop clover hay.

Ration No. III
1.5 lbs. of second crop clover hay, .1 lb. of corn fodder, and .3 lb. of oats and corn, equal parts.

SHEEP

Ration No. IV
1.8 lbs. of second crop clover hay, 1.5 lbs. of roots, and .3 lb. of shelled corn.

Ration No. V
2.6 lbs. of fodder corn, 1.5 lbs. of roots, and .3 lb. of oats and shelled corn, equal parts.

Ration No. VI
2 lbs. of oat straw, 1.6 lbs. of roots, .6 lb. of bran and oats, equal parts.

How to Feed Above Rations.—A glance at the above rations gives one an idea that it would be impractical to weigh out feed so carefully to each sheep; which conclusion of course, is true. To feed any of the above rations, one would simply need to know the number of ewes to be fed, and their approximate weight. (The average ewe will weigh between 125 and 150 lbs.) If one had 40 ewes weighing 140 lbs. each, he would have 5,600 lbs. of sheep. If each 100 lbs. of sheep required 3.7 lbs. of clover hay, his flock would require 56 x 3.7 lbs. or 207.2 lbs. of clover hay per day; and about one half of this amount would be scattered in the feeding racks each morning, the balance in the evening.

If one will weigh a few forkfuls of hay occasionally, he can tell very closely, without weighing every time he feeds, about how much hay is fed each time.

If one is to feed a mixture of corn and oats, equal parts, he would simply mix together one or more hundred pounds of each; then, by weighing a few measurefuls of the mixture, he can tell approximately the right amount of grain to feed to his flock without weighing the grain each time he feeds.

Questions:
1. Why do breeding ewes require food during the winter?
2. What are the results of overfeeding breeding ewes? Of underfeeding?
3. How would you proceed to feed a flock of ewes approximately the right amount of feed?

Arithmetic:
1. How much will it cost to feed a 140-lb. ewe 200 days, on ration No. I, if fodder corn is worth $5.00 per ton?
2. How much will it cost to feed a 140-lb. ewe 200 days on ration No. IV, if clover hay is worth $5.00 per ton, roots $2.00 per ton, and corn and oats $20.00 per ton?
3. How much will it cost to feed a 140-lb. ewe 200 days, on ration No. VI, if straw is worth nothing, roots are worth $2.00 per ton, and oats and bran are worth $24.00 per ton?

CHAPTER XVIII

SWINE

TYPES AND BREEDS

Types.—Swine are bred entirely for pork, consequently there is not so much difference in breeds of hogs as in breeds of other stock. There are two general types of hogs, namely, lard hogs and bacon hogs.

Lard hogs are by far the most common in America. They have short legs, comparatively short broad backs and heavy hams and shoulders. They have been bred and fed to mature early and fatten easily. The most common breeds of lard hogs are Poland China, Berkshire, Duroc-Jersey, Chester White and Small Yorkshire. Other breeds in this class are Hampshire, Essex and Suffolk. Poland China is an American breed, black in color with six white points, a white spot in the face, white on the end of the tail and four white feet.

Figure 100.—Berkshire hog, a lard type.

The ears droop. Berkshires are an English breed. They are black and have the six white points the same as the Poland China. Their ears stand erect or point outward. They are not quite so wide and have a little longer legs than Poland Chinas. Duroc-Jerseys were developed in the United States. They are very much like the Poland China in every way except that they are red in color. Chester White is another United States breed. They are larger than the Poland Chinas. In form they are quite like the Poland Chinas, but are white in color. Small Yorkshires originated in England. This is one of the smallest breeds of hogs. They are white in color and have very short heads with a face very much dished, that is, the nose has the appearance of being broken and turned up. The ears stand erect. The Hampshire is an American breed, black with a white belt about the body. Sometimes it is classed as an intermediate breed. Essex is a small black English breed.

Intermediate Breeds.—There are three breeds that can hardly be classed as either lard or bacon in type. They are partly both. They are all white, and medium in size.

Figure 101.—Improved Yorkshire hogs, a bacon type.

The Cheshire is a United States breed, with ears erect. The Victoria is a United States breed, with ears drooping. The Middle Yorkshire is an English breed, with ears erect and face slightly dished.

Bacon breeds are large, long hogs, with very deep bodies, long legs and long heads. They are not so broad as the lard hogs and the hams and shoulders are lighter. These breeds have a tendency to mix more lean with the fat than the lard hogs. Bacon with nice strips of lean running through it is to be desired. There are but two breeds in this group, the Yorkshire and the Tamworth. These are both English breeds. The Yorkshire is white, with dished face and erect ears. The Tamworth is red, with a very long nose and ears erect.

Questions:
1. What are the chief differences between bacon and lard hogs?
2. Describe four important breeds of lard hogs.
3. What breed is most commonly raised in your community? Describe this breed fully?

Arithemtic:
1. If a farmer raises 60 hogs that weigh 225 lbs. each, and sells them at 7½c. per lb., how much money will he get?
2. A farmer turns 8 50-lb. pigs into an acre of clover pasture and leaves them 70 days. In the meantime he feeds them 1,000 lbs. of shorts, worth $25 per ton. At the end of the period each pig weighs 100 lbs. How many pounds did the pigs gain? How much is the amount they gained worth at 7½c. per pound? How much did the farmer make, after paying for the shorts?

CARE AND MANAGEMENT
THE SWINE INDUSTRY

Profitable Meat Production.—Hogs are kept on nearly every farm, but only a small proportion of farmers raise enough hogs to make pork production an important factor in the income of the farm. Pork production is, however, a very important enterprise on many farms, and has in many cases proved profitable; in fact more profitable than any other kind of meat production.

Advantage of Hog Raising.—Some advantages of pork production over that of other kinds of meat are:

(a) A brood sow may produce from four to twenty pigs in a year. On this account the cost of a pig at birth is less in proportion than the cost of a calf or a lamb.

(b) The fact that hogs have large litters, reach maturity quickly and do not require expensive shelter, enables one to get started in raising hogs more quickly and with less expense than is required to start with other kinds of live stock.

(c) Less labor is required to care for hogs than to care for enough cattle to bring in the same amount of money.

(d) They consume and convert into valuable products the wastes and slops of the farm.

Disadvantages of Hog Raising.—The main disadvantages of hog raising are:

(a) Hogs are not able to use the coarse roughage, as corn stover and straw, that is usually found on the farm; hence cannot convert these products into salable form as can sheep and cattle.

(b) Their chief feed must be grain, at least for fattening, and grain feed is more expensive than roughage.

(c) They are more likely to be taken off in large numbers by disease than are the other classes of live stock.

Possibilities.—Hogs have probably been the means of paying off more mortgages than has any other class of stock. A young man wishing to make a start on the farm can well afford to give careful attention to hogs and their possibilities of producing a profit.

Figure 102.—Some good porkers of the bacon type.

A good brood sow should have from six to ten pigs at a litter, and, if desired, may have two litters a year. Pigs, when eight months old, should weigh 200 lbs., or more. If a sow produces fourteen pigs in a year, and each pig when eight months old weighs 200 lbs., she would produce 2,800 pounds of pork in a year, which as 5c. per pound would be worth $140.

Hog Cholera.—The most dangerous disease of hogs is hog cholera, and it has caused the loss of millions of dollars' worth of hogs in the United States. Veterinarians have now discovered a method by which it is possible to vaccinate hogs and prevent their having cholera. They vaccinate in much the same manner as people are vaccinated to make them immune to small-pox. Vaccination is quite expensive and proper facilities are not always available; so it is well to take every precaution to prevent the disease.

Hog cholera is a contagious disease. That is, hogs are very likely to take the disease if they come in contact with other hogs that are infected with it. The germs may be carried from one pen to another, or from one farm to another, on one's cloths, by dogs, by running water or by any other method by which particles of dust or disease germs might be carried about.

Preventive Measures.—If hogs are kept in clean, healthful quarters, given plenty of exercise, and fed, except when fattening, enough muscle-forming food, as clover pasture, clover hay, milk, shorts, etc., to keep them in good, vigorous condition, they will be better able to resist the disease than if they are kept in less thrifty condition. If cholera breaks out in the community, one should use every precaution to prevent the germs from being brought on the farm; and, if it gets very close, it is well to dispose of all the hogs that are well and fit to sell. Chances of loss may be greatly reduced by separating the hogs.

Questions:
1. What are some of the advantages of pork production over the production of other classes of meat? What are some of the disadvantages?
2. What can you say of the possibilities of pork production?
3. Tell all you can about hog cholera.

Arithmetic:
1. What is the value of a hog weighing 225 lbs. at 5½c. per pound?
2. A sow has 7 pigs in a litter. When 8 months old the pigs weigh 200 lbs. each. What is the weight of all? How much are they worth at 5½c. per pound?
3. If a bushel of corn will produce 10 lbs. of pork, how much will the feed for the production of a pound of pork cost, if corn is worth 35c. per bushel?

THE BROOD SOW AND PIGS

The brood sow and her care and feed determine the cost of pigs at birth. In the first place a sow of good type, and of the breed desired, should be selected. If several litters of pigs are raised, it is well to have some method of marking the young pigs, so that when they are grown one can tell from which litter they came. It is desirable to have brood sows that will have large litters of pigs; and if one selects brood sows from a large litter one is more likely to get a good number of pigs from each sow than if sows were selected from a bunch of hogs without regard to whether they came from large litters or not.

If one does not mark the pigs at birth, young sows may be selected from small litters, because the sows with small litters feed their pigs a little better, and as a consequence the pigs are usually a little fatter and better looking than the pigs from large litters.

Care of the Brood Sows.—The brood sow should have plenty of succulent and muscle-forming feed, but should

Figure 103.—A sow with a large litter. An essential of cheap pork production is a low cost of pigs at birth.

not be overfed. She should have at all times plenty of exercise. It is a mistake to allow brood sows to run during the fall with the hogs that are being fattened. It is a waste of feed, and the sows are injured if allowed to get too fat. During the fall the brood sows should have the run of a good pasture, with only enough grain to keep them in good thrifty condition.

Shelter.—If only one litter of pigs is to be raised from each sow each year, it is well to have them come as early in the spring as the weather is warm. Then no expensive shelter is needed. A small cot (movable house) well banked, or a straw shed, is ample for the sow during the winter and in summer all that is needed is shelter to keep the pigs dry and to protect them from the sun. If cots are used they may be moved to the pasture for summer shelter. Hogs need shade in summer.

If one is going to raise two litters of pigs from a brood sow in a year, good warm quarters must be provided. These quarters need not be expensive, but they should be convenient and comfortable.

Requirements for Pigs. — The first requirement of young pigs is that they have a clean, dry, comfortable bed in which to arrive. As the mother is naturally in a feverish condition at this time, she may be somewhat careless and lie on the little pigs. To prevent this, a shelf ten or twelve inches wide and eight or ten inches from the floor should be built around the pen, so as to make room for the pigs to get out of the way of their mother. This is a very simple precaution, and may save a number of pigs.

Figure 104.—Hog cot, a cheap and portable shelter.

A Creep.—To feed the small pigs so that the sow cannot bother them, have a small yard or pen fenced off in the place in which the sow is kept, with the fence raised high enough from the ground so that the little pigs can pass back and forth easily, but low enough to keep out the old sow. In this place plenty of trough room should be provided, so that every pig has a chance to eat. Otherwise, the larger, stronger pigs will get most of the feed and the smaller ones will not get enough.

Weaning Pigs.—If but one litter of pigs is raised per year, they may be allowed to run with their mother until from twelve to sixteen weeks old, or even longer, until the sow begins to wean the pigs herself. If the young pigs are given a chance to learn to eat as suggested above, they may be weaned with very little difficulty at any time after they are six weeks old. If the sow is doing well, it is usually better to leave the pigs with her until they are about twelve weeks old.

Comfort.—In fattening any kind of stock, comfort is an important factor, and one who overlooks it is a loser thereby. If hogs are fed in the field, a good soft and dry bed should be provided for them or they will not do their best, and they should always have a supply of fresh water. The same is true if they are kept in the yard. Some feeders claim, and with good reason too, that an armful of straw may often be as valuable to a bunch of hogs as a bushel of corn.

Fencing is the most expensive part of furnishing pasture for hogs, but as a rule it is cheaper than the labor of caring for and carrying feed to the hogs would be. The cost of fencing may be reduced by having comparatively large, well-shaped fields, and by planning for them a rotation that will furnish the maximum amount of feed. A four-year rotation, of (1) grain, (2) clover, (3) and (4) corn, on four fenced fields of uniform size, is very satisfactory. One of the four fields would be in grain, one in clover and two in corn, each year; the clover and the two corn crops to be fed off by the hogs.

Arrangement of Fields.—Four fields adjoining the farmstead, each ½ acre to one acre in size, for each brood sow kept on the farm, make it possible to produce both summer and fall feed for hogs very cheaply. Each year one field would be sown to grain and red clover seed, and another field would be in pasture, and the other two in corn. Such rotation once established would supply abundance of cheap feed with the least labor.

Questions:
1. What are some of the points worth considering in the selection of a brood sow?
2. What can you say concerning the care and shelter of brood sows?
3. How may young pigs be fed so they will not be bothered by their mother?
4. What is the best kind of summer feed for hogs, and how supplied?
5. In what way can the cost of fencing for hogs be reduced?

Arithmetic:

1. If ½ acre of clover pasture is required for a sow and eight pigs, how many acres are required for six sows with litters?

2. How many acres of land in a field 20 rods by 24 rods in size? How many rods of fencing are required to enclose it? How many rods of fencing per acre? What would the fencing cost per acre at 50c. per rod of fencing?

3. If it costs $15.00 per acre to fence a field, what is the annual cost of the fence, if it lasts ten years and interest is charged at the rate of 4%? (Ans.: 1-10 of $15.00 and interest on $15.00.)

FEEDING

FATTENING HOGS ECONOMICALLY

To fatten hogs it is simply necessary to supply them with plenty of food, as they usually have a good appetite and are

Figure 105.—Hogs helping themselves to the corn crop.

not easily injured by overfeeding. It is wise, however, to change their feed to another gradually, that is, where they are being fed all they will take.

The majority of hogs are fattened in the fall and early winter; and on that account we will suggest some of the better methods practiced in fall fattening.

Labor.—A very common practice followed in fattening hogs is to shut them up in a small yard and feed them generously. This practice, however, is no longer regarded as desirable. The animals are not kept in the most rugged condition by being confined too closely; they are more likely to become diseased, and a great deal of labor is necessary to feed and care for them when closely confined. Labor is one of the very important items in the cost of pork production, and every effort should be made to reduce the necessary labor to the minimum.

Early Fall Feed.—During the early part of the fattening season considerable green and succulent feed can be fed to advantage. This feed is very easily supplied by raising a patch of pumpkins near the hog pasture, where the pumpkins can be easily thrown over the fence to the hogs. Plenty of good pasture is also desirable at this time, and it may be supplied in any way most convenient.

Field Peas.—In the first part of the fattening period the pigs will make considerable growth; so some muscle-forming feed is desirable as a part of the ration, rather than an exclusive corn diet. A small field of field peas, so situated that they may be harvested by the hogs when ripe by turning the hogs into the field, gives the hogs an excellent start at slight expenditure of labor. The peas are sown very early in the spring, at the rate of three bushels of seed per acre, and nothing more is done to them until the hogs are turned in.

Corn.—After the peas have been fed off, or as soon as the corn is ripe, if one has no peas, the hogs may be turned into a portion of the cornfield and allowed to help themselves. If some green feed is provided by sowing rye or rape in the corn, at the time it was cultivated last, a large amount of green feed will be supplied at very small cost, and will be relished by the hogs along with the corn.

It costs only about half as much to grow a crop of corn up to the time it is ripe, as to raise, cut, husk and feed it. In other words, it costs about $5.00 per acre to cut, shock and husk one acre of corn, and by allowing the hogs to harvest the crop themselves this cost is saved. Of course the field must be fenced; but, if regular fields are provided near the house, on which to raise pasture, peas and corn for the hogs, and these fields are permanently fenced, the annual cost of fencing is not a very large item.

Not all the corn raised would be fed off, for it is not well to have the hogs in the field after snow and cold weather come. Six or eight pigs five to seven months old will ordinarily clean up an acre of average corn during the fall.

Waste by Hogging Crops.—Allowing hogs to help themselves to a crop is called "hogging off the crop." This practice is generally regarded as very wasteful—that the hogs trample down and waste a great deal of the crop.

Figure 106.—Hogs in rape, a good pasture crop. The seed is cheap and may be sown six to eight weeks before the pasture is needed.

There is, however, very little of the crop wasted if the hogs are turned into a small patch (sufficient to last them two or three weeks) at a time.

Results obtained by the Minnesota Experiment Station, and by many practical farmers, show that an acre of corn will make fully as much and often more pork, where hogs help themselves, than where the corn is husked and fed to the hogs in a yard.

At the Minnesota Experiment Station one lot of hogs was turned into a field of corn, and another similar lot was shut in a yard and fed husked corn. It was found that the hogs in the field required 7.35 lbs. of grain to make a pound of gain, while those in the yard required 8.59 lbs. of grain to make a pound of gain.

Questions:
1. In what ways may we reduce the amount of labor necessary in caring for fattening hogs?
2. Of what use are field peas as a feed for hogs?
3. What can you say regarding "hogging off corn"?

SWINE

Arithmetic:

1. If it costs $1.25 to plow an acre of land, 50c. to harrow it 3 times, 25c. to plant it, $1.60 to cultivate it 4 times, and 60c. to manure it, 25c. for seed, and $3.00 for rent, how much does it cost to raise an acre of corn?

2. If it costs $7.50 to raise an acre of corn, and $2.00 annually to fence it, how much does it cost per acre? How much does the corn cost per bushel, if it yields 40 bushels per acre?

3. If it requires 8 lbs. of ear corn to make one pound of pork, how many pounds of pork will 40 bus. of ear corn make? (72 lbs. per bu.) How much will the pork made from 40 bus. of corn be worth at 5c. per lb.?

FEEDING SOWS AND PIGS

Feeding the Brood Sow.—It is very easy to overfeed a brood sow in winter. If she has raised two litters of pigs during the year, so she is likely to be thin in the fall, she will need considerable feed until she begins to fatten up a little. If she has raised but one litter, which is the practice on most farms, she will have had the whole fall to fatten up, and very little grain is necessary or desirable during the winter.

Bulky Feed.—If only one or two sows are kept, the slops from the house furnish an excellent form of bulky feed, which helps to satisfy their appetites, but really contains little nutriment. If a large number of sows are kept, the slops from the house do not go very far, and one is likely to feed them more grain, to keep them from squealing, than they really need. If supplied with good clover hay, hogs will soon learn to eat it; and this furnishes the bulk

Figure 107.—Pigs in clover, the best and cheapest summer feed for pigs.

they need and some nourishment, so that they do not need so much grain to satisfy them. Roots are an excellent form of feed for brood sows. As they are succulent, they aid in digestion, supply bulk and variety, and tone up the system.

Suggested Rations.—The following grain mixtures have been fed to brood sows, in addition to clover hay, with very satisfactory results:

1. Shorts 1 part, corn 3 parts (by weight).
2. Oil cake, 1 part, corn 7 parts (by weight).

About ¾ lb. of either of these mixtures per day, per 100 lbs. live weight of hog, is sufficient, if enough bulky food, as hay or roots, is fed to satisfy the appetite.

Mother's Milk for Young Pigs.—The very best feed for young pigs for the first few weeks is the dam's milk. If the sow has been well cared for previous to farrowing and is liberally fed after farrowing, she will, if she is the right kind of a mother, give a liberal amount of milk. At two or three weeks of age the young pigs begin to develop a desire for something besides their mother's milk. Provision should be made to feed them some light but muscle-forming food, as skimmed milk with a little meal added.

Clean Feed.—Only clean, wholesome feed should be fed to the small pigs, and the trough in which they are fed should be kept clean; because their digestion is easily deranged and a pig is valuable only when his digestion is good. No feed should be left in the trough from one feeding time to the next. Feed only what they can eat up clean.

Keep Pigs Growing.—The aim in feeding young pigs should be to keep them growing every day; and, since their capacity to make use of feed determines their usefulness, it is well to so feed them as to strengthen and develop this capacity. Bulky feeds containing a good proportion of muscle-forming feed, as milk, milk and shorts, clover pasture, etc., are very good kinds of feed for young pigs.

Summer Feed.—If but one litter of pigs is raised per year, they should come in the spring and can be raised on pasture. Pasture furnishes the cheapest feed on the farm, and good feed too. Pigs are often kept in small pens and fed grain and slops. This practice is undesirable for several reasons. It is expensive both in labor and in feed. It does not provide the exercise necessary for the best development of growing pigs or breeding stock. Last, but not least, it is difficult to find any feed so well adapted to the growing pig as good clover pasture, supplemented with milk from the sow, skimmed milk and good, clean slops thickened with shorts or other muscle-forming feed.

Pasture.—Red clover furnishes very cheap pasture, because the seed is sown with the preceding grain crop and no plowing or preparation of the land is necessary. Hogs relish young and succulent pasture; and as clover grows very rapidly during the early part of the summer, and the young pigs do not eat as much as they will later, the clover, if the pasture is large enough, usually gets ahead of them. It is well, then, to cut with a mower a small strip of the clover next to the pens, early in June. This part will start up soon and furnish the best kind of pasture. The rest may be cut for hay the latter part of June. The second crop will come on, and the hogs will be larger and need more feed than earlier, and will likely keep pace with the growing clover.

Rape, rye, field peas or any other of the grain crops furnish good annual pasture for hogs, if for any reason one has not the clover. Blue grass and white clover, or bromus and white clover, make very good permanent pastures, if it seems undesirable to rotate the crops and thus supply clover pasture.

Questions:
 1. Why is one likely to overfeed brood sows?
 2. In what way may the tendency to overfeed brood sows be overcome?
 3. How much grain does a brood sow need per day per 100 lbs. live weight?

Arithmetic:
 1. If a sow weighs 350 lbs. and requires ¾ lb. of grain per day per 100 lbs. live weight, how much grain should she receive per day?
 2. If it costs $10 per year to keep a brood sow, what is the average cost per pig at birth, if she raises five pigs? If she raises eight pigs?
 3. If 100 lbs. of shorts worth $20 per ton and 300 lbs. of corn worth 42c. per bushel (56 lbs.) are mixed together, how many pounds of feed will there be in the mixture? What is the average price per pound? If a 400-lb. sow is fed ¾ lb. of the mixture per 100 lbs. live weight, how much will her grain ration cost per day?
 4. What are three important points in the feeding of little pigs?

CHAPTER XIX

POULTRY, BIRDS AND BEES

POULTRY ON THE FARM

Importance of Poultry Industry.—Many persons believe that raising poultry is a small business, hardly worthy of a man's time, and that a few chickens are good simply as a

Figure 108.—A neat flock of Barred Plymouth Rocks.

pastime for those who live in town, or as a source of pin money for women on farms.

It is true that the greater part of the poultry raised is raised in just this way; but, in spite of this fact, poultry brings to the farms of the country many millions of dollars annually. In fact, the poultry product of the United States is greater than the dairy product. In many dairy communities, where creameries ship out twenty to fifty thousand dollars worth of butter a year, the poultry and eggs sold bring in as much, and in some cases more, than the butter.

Poultry Records.—The Minnesota Experiment Station has been gathering very accurate statistics for several years, on eight average farms in each of three counties in Minnesota, namely, at Halstad, Norman Co.; Marshall, Lyon Co.; and Northfield, Rice Co. These statistics show that the value of poultry products, used and sold per farm in 1908, was $56.61 at Halstad, $95.75 at Marshall, and $150.43 at Northfield. This makes an average of about $100 per farm for poultry products per year.

Figure 109.—A White Leghorn cockerel and hen, a type of lighter breeds, which are the better layers.

In the United States in 1909 there were produced 1,591,311,371 dozens of eggs on farms. The value was $306,688,960. Practically the same farms reported the production of 445,650,124 fowls, valued at $202,506,272.

A Start with Poultry.—Boys and girls who feel a liking for poultry are urged to undertake it as a means of making a little spending money, of earning their way through school, or as a business worthy of study. The mother should be relieved of a part or all of the care of the poultry. She no doubt will share the earnings with them very liberally. If poultry is really a part of the farm business, the father will, or should, at least, be glad to give them a definite part of the work to do and a share in the income, or preferably give them entire charge of a portion of the flock. If there is a place where poultry may be kept, a few specimens of some preferred breed should be secured as a start. There will be many things to learn, but information gained in this way is fully as important as what is learned at school. Father and mother will be able to assist; and, if there is a poultry man in the vicinity who is doing well, he will no doubt give a great deal of valuable information. A few

Figure 110.—A White Wyandotte cockerel and hen, a type of the heavier breeds valuable for meat production.

poultry bulletins and papers should be read, and interest and success are almost sure to follow.

Breeds of Poultry.—There are a great many breeds of poultry, and most of them are good under special conditions. There is no best breed. If no particular breed is preferred, good specimens of a breed common in the neighborhood should be secured.

All common breeds of chickens may be divided into three classes—egg, meat and general purpose—according to what they are adapted to produce, just as cattle are divided into dairy, beef and general purpose classes.

In the egg producing class we have the Leghorns, Minorcas, Spanish and Andalusians. In the meat producing class are found the Cochins, Brahmas, and Langshans. And in the general purpose class, or those well adapted to produce both eggs and meat, are the Plymouth Rocks, Wyandottes, Rhode Island Reds, and Orpingtons.

There are good and poor birds in any breed, and the

Figure 111.—A Barred Plymouth Rock cockerel and hen, a type of the heavier breeds.

only way to be reasonably sure of getting good chickens is to get them from a flock that has a good record as producers of either eggs or meat, or both.

Questions:

1. Compare the poultry and the dairy industries.
2. Does the poultry in your neighborhood receive as much attention as the dairy?
3. What is the conservative estimate of the poultry product of the state of Minnesota?
4. What are the three classes into which all the common breeds of poultry may be divided?

Arithmetic:

1. If a farmer keeps 50 hens and each hen lays 125 eggs in a year, how many dozen eggs will the farmer get in a year? How much will these eggs be worth at 20c. per dozen?

2. If a farmer keeps 50 hens and half of them produce 10 chickens each, how many young chickens will he have? How much are they worth at 35c. each?

3. If 9,000 lbs. of grain, worth 1c. per pound, is required to keep 50 hens one year and raise 250 chickens, what is the total cost of feed?

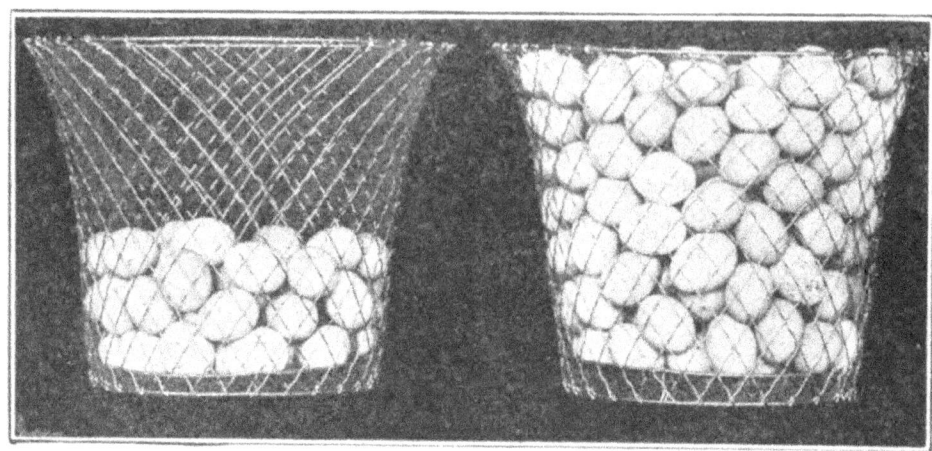

Figure 112.—Two baskets of eggs. The one on the left represents the 75 eggs laid by the average hen in one year. The one on the right represents 220 eggs laid in a year by the best hen at the Crookston Experiment Station.

CARE OF POULTRY

Sitting Hens.—Like any other class of live stock, poultry, to do well, must be well cared for. During the early part of the summer the young chicks require considerable attention. A great deal of time may be saved, if several hens are set at one time in a building separate from the main poultry house. We mention raising chicks in this way, because but comparatively few persons use incubators. A woodshed or corncrib that is clean, and that may be darkened, is a good place in which to set hens. In the evening take as many broody hens as you can get, or as you want, and put them in this shed or crib, that has been well cleaned and provided with good, clean nests, preferably near or on the floor. Shut these hens in over night and darken the windows. If they continue broody the next morning, set them at once with eggs selected from the best hens. Then provide them with plenty of shelled corn or other grain, fresh water and a box of ashes or road dust for their dust baths. It will be well also to dust some insect powder in the nest to keep away lice and mites. It is but little more work to care for a dozen sitting hens in this way than to care for one.

The Young Chicks.—For the first day or so after the young chicks are hatched they will not need anything to eat, and it is well to keep them in the nest. If the room is darkened, the old hens will not be in such a hurry to leave the nests. When the chicks are about thirty or thirty-six hours old, give them some bread crumbs slightly moistened in milk. Feed them several times during the day. After a day or so, some ground oats, with the hulls removed, may be added, and after a week or ten days some ground or cracked grain, as corn or wheat, may supply a part of their ration. Very small kernels of wheat and millet seed are also very good. The chicks should be supplied at all times with pure, fresh water and fine grit. Too much care cannot be taken in keeping their water and feed clean. Plenty of exercise is also necessary. On the farm the chickens usually have the run of the whole place, which is the best possible condition for them, as they can then get exercise, insects, grit, and green food—things that are not so easily supplied when they are confined.

Care of Hens in Winter.—Eggs are one of the chief products of poultry, and one's success in the business usually depends upon getting eggs in the winter, when they bring a good price. Pullets hatched early in the spring are more likely to lay during the winter than old hens. To get eggs in the winter, one must supply as nearly as possible summer conditions. In other words, chickens must be forced to get exercise by scratching for their feed, as is necessary, if their grain feed is thrown in loose straw or litter. They should have something to take the place of bugs and worms that they get in the summer. Scraps of meat and ground bone will answer. They must have something to take the place of the sharp stones and gravel that they pick up as they run about the fields. Crushed stone or crockery will supply this grit. Such material is sold on the market as grit, and poultry should always have a supply in winter. The grit aids them in digesting their food. They need something to take the place of the green food they get in summer. Cabbages, beets, potatoes or sprouted grains will supply this need. Some material containing lime, from which they can make egg shells, is also necessary. Crushed oyster-shells, kept constantly before them, will supply the necessary lime. They must be kept comfortable—that is, their house should be kept warm enough so that their combs will not freeze. Their houses should be sufficiently ventilated to supply fresh air and keep the coop dry. They should have a supply of pure water, a place for a dust bath, and a clean coop free from vermin. If these few simple precautions are observed and a liberal supply of a variety of grains, as wheat, barley, oats, and corn, with an occasional mash, are supplied, poultry should prove profitable.

Questions:
1. In what way may the work of caring for sitting hens be lessened?
2. What can you say regarding the care of young chicks?
3. What conditions must be provided for hens in winter, if they are to lay?

Arithmetic:
1. The average hen lays about 75 eggs per year. What are the eggs worth at 20c. per dozen?
2. Some hens lay 200 eggs per year. What are the eggs worth at 20c. per dozen?
3. If a hen can cover 15 eggs, how many eggs will 7 hens cover? What would these hens cost at 50c. each?
4. If each hen lays 100 eggs, worth 20c. per dozen, and raises 10 chicks, worth 30c. each, in a year, what is the annual income per hen? What is the total annual income from 50 hens?

A 100 HEN POULTRY HOUSE

A Poultry House.—There are a great many types of poultry houses which include all the principles required. It is simply a matter of choice with the owner. We describe a very common type, merely to emphasize the essentials.

We will explain in detail a house 16 x 32 feet in size, large enough for from eighty to one hundred hens.

It is placed on well drained land, somewhat protected from the north and west, and stands the long way east and west, with the high side to the south.

The foundation is of stone or concrete, set in the ground at least one foot, and extending above ground six inches. A sill 4 x 6 inches is placed on top of the foundation and the studding spiked on top of the sill. The house is 4½ feet high at the back and 8 feet high at the front. It has a shed roof made of boards covered with prepared roofing. The studdings are placed two feet apart and boarded with rough boards. The building is then papered with building paper and sided.

Doors and Windows.—A door is placed in each end, near the south side; and four windows, about 2 x 4 feet in size, are made in the south side. They are placed high, so the sun will shine clear to the back part of the coop. There will be an opening near the floor on the south side, through which the hens may be let out, or a door may be placed in this side, if desired. The windows are made to slip up and down, the same as in a house. At least one of the windows is provided with a muslin or duck curtain; and, except in the most severe weather, the upper sash is let down and the opening covered with the canvas. The canvas may be on a frame, hinged at the top, or simply tacked in the opening. This provides ample ventilation without draught and keeps the air pure and dry. In cold weather the windows may be partly closed, but never entirely.

Inside Finish.—The inside may be left with the bare studding or preferably ceiled with matched lumber. The house is partitioned into two parts, the lower three feet of the partition being of boards, so that the fowls cannot fight, and the upper part of wire netting.

Roosts.—To make it easy to clean the coop, and to leave all the floor space available for feeding and exercising, a platform three feet wide is built against the north wall 2½ feet from the floor. This is to catch the droppings and should be made of matched lumber. Arms of 2 x 4's, 2½ feet long, are attached to the back wall one foot above the platform, to extend out over the platform. Legs are placed under the outer ends of these arms to hold them level with the platform. On top of these arms and at right angles to them are placed two poles, or 2 x 3's edgewise, with corners rounded off, for roosts. The back roost is about one foot from the wall and the second fifteen or sixteen inches from the first. A muslin curtain is hung from the ceiling to drop in front of the roosts when needed. To clean the droppings from the platform the roosts may be unhooked

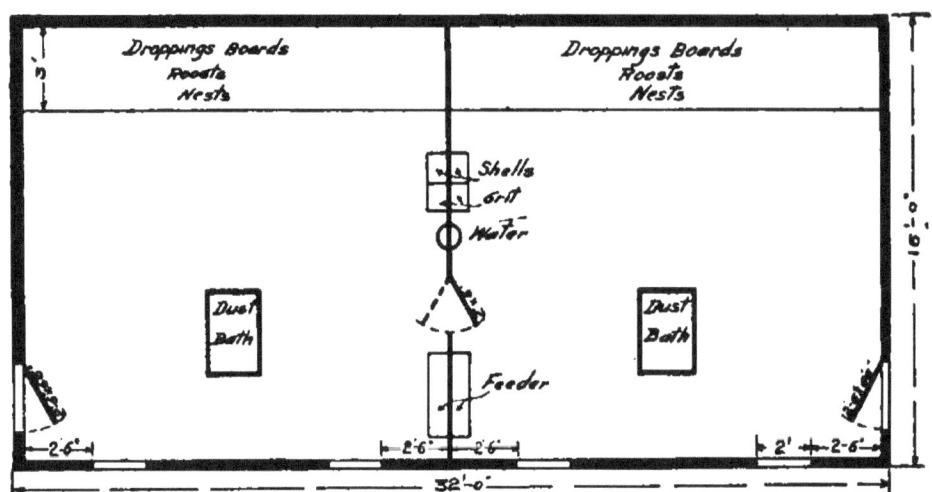

Figure 113.—Floor plan of poultry house.

and removed, or the front side raised, and both hooked to the ceiling. They are then out of the way, so that the cleaning may be thoroughly done.

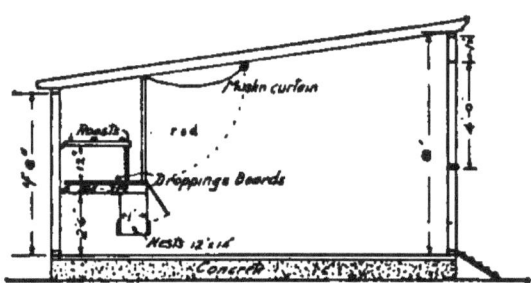

Figure 114.—Cross section of poultry house. Note arrangement of roosts and nests leaving floor space open.

Nests.—For nests a box 8 to 12 feet long and 12 inches wide is made, with sides 8 inches high. This is divided into nests 12 to 14 inches long, with partitions 18 inches high. Hens prefer nests that are rather dark; so a good place for them is under the front edge of the platform. Place this row of nests close up under the front edge used for a droppings board. Then hinge a 10-inch board in front of the nests to close the 10-inch space between the platform and the 8-inch board on the front side of the nests. This darkens them somewhat. The hens enter from behind. The high partitions between the nests prevent the hens from going from one nest to the next, so they are not likely to disturb one another and break or dirty eggs. When gathering the eggs the board in front is opened. These nests should be loose, so that they may be taken out to be aired, sunned and sprayed, when desired, to keep them free from mites and lice.

The Floor.—A cement floor covered with a little sand and clean litter is the best for a poultry house, as it can be kept dry and clean with very little work.

With a poultry house as described, one can keep poultry clean and comfortable the whole year through.

Questions:
1. What can you say regarding the size and arrangement of windows in a poultry house?
2. What advantages are there in having the roosts and other inside fixtures of a poultry house removable?
3. Describe a good arrangement for nests in a poultry house.

Arithmetic:
1. How many yards of muslin in a strip 16 ft. long and 3 ft. wide? How much would it cost at 12c. per yard?
2. How many feet of lumber would be required to build a platform 3 ft. wide by 16 ft. long? (Allow 1-6 for matched lumber.) How much would it cost at $30.00 per thousand feet?
3. To make a row of nests 12 ft. long, as described above, requires 2 pieces 1 x 8, 1 piece 1 x 10, and 1 piece 1 x 12—all 12 ft. long, and 1 piece 1 x 12, 20 ft. long. How many feet of lumber are required? How much would it cost at $24.00 per thousand feet?

FEEDING LAYING HENS

Rations for Laying Hens.—It is not easy to determine the amount of feed eaten by an average laying hen. If a person is to study the poultry business from a practical point of view he must know approximately the requirements of his flock. As stated in the last lesson, food requirements vary; but it is well to have some basis for one's calculations. We submit a few rations that have been fed by practical poultrymen, successfully. The following rations are based on the requirements of an average laying hen for one day.

Ration No. 1, reported by The Cornell Experiment Station. See Bulletin No. 212, page 11: .175 lbs. wheat; .07 lb. ground bone; .022 lb. cabbage.

16—

The above is the average amount fed per day to each hen. The actual amount varied each day as the requirements of the hens varied. Early in the morning a part of the wheat was scattered in the straw on the floor of the coop. In the middle of the afternoon the flock was fed all the cooked cut meat and bone they would eat up quickly. Then at 4:30 p. m. they were given a liberal feed of wheat again. The wheat was always scattered in the straw to make the fowls exercise. The cabbage was fed by having a head suspended from a wire and in reach of the hens.

In addition to the feed, the hens always had a supply of fresh water, grit and oyster-shells. Grit is necessary to aid in the digestion of the food, and oyster-shells are necessary to supply the material of which egg-shells are made.

Ration No. 2, reported by the Utah Experiment Station, Bulletin No. 92, page 139: .156 lb. mixed grain; .022 lb. ground bone; .004 lb. beef scraps; .008 lb. gluten meal; .066 lb. skimmed milk.

As in ration No. 1, grit, shells and water were supplied in addition to the feed. The figures represented the average amount of each kind of feed given per day to each hen. Ration No. 2 supplies a little more variety than No. 1, but has the disadvantage of necessitating the purchase of feeds not raised on the farm.

Variety.—Hens, like any other class of stock, like a variety of feeds, and in this respect the last ration is good. Variety in grains is very desirable, and is easily provided by feeding a mixture of the common farm grains. Corn can profitably be added to the ration, even of laying hens, to replace a part of the grain feed. Corn is the cheapest of the grain feeds. If corn is fed, it is well to give it the last thing in the afternoon, so that the hens may have a crop full when they go to roost; but a large amount of corn is too fattening. Millet seed, buckwheat and sunflower seed are good feeds to add in small amounts to the grain ration, to add variety.

Green Feed.—Chickens will eat quite an amount of fine cut clover or alfalfa, or the leaves that may usually be gathered in the mow or at the bottom of the chute where clover or alfalfa hay is thrown down. This may be fed dry, or moistened and mixed with a little ground feed. Hens will eat several pounds of this material during the winter, if given the opportunity.

Mangels, beets or carrots are also valuable feeds for laying hens in winter. Such feeds take the place of the green grass hens eat in the summer.

Ration No. 3.—Fed with good results by the Maine Experiment Station. See Farmer's Bulletin No. 357, page 33, U. S. Dept. of Agriculture: .246 lb. grain and meal; .011 lb. oyster shell; .006 lb. dry cracked bone; .005 lb. grit; .006 lb. charcoal; .027 lb. clover.

The above is the average amount given per hen per day. This ration was fed in practically the same manner as ration No. 1, except for the mash or ground feed. This mash was made of a mixture of 2 parts by weight of bran, 1 part cornmeal, 1 part middlings, 1 part gluten meal, 1 part linseed meal and 1 part beef scraps, and was kept constantly before the birds in a feed trough with a slatted front. This mash was fed dry.

A simpler mash, composed of bran, some ground grain of any or several kinds, and, if possible, a little beef scrap, would likely do as well as the more complicated mash.

Questions:
1. Describe the method followed in feeding ration No. 1.
2. What can you say regarding feeding hens several different kinds of feed?
3. In what ways may green feed be supplied to hens in winter?
4. Describe the composition and method of feeding the mash in ration No. 3.

Arithmetic:
1. How much of the various kinds of feed will a hen eat in 200 days, if fed ration No. 1?
2. How much will it cost to feed a hen 200 days on ration No. 1, if wheat is worth 85c. per bushel, ground bone 1c. per pound, and cabbage $5.00 per ton?
3. How many eggs must a hen lay in 200 days to pay for her feed, if she is fed ration No. 1 and eggs are worth 25c. per dozen?

BIRDS

Importance of Birds.—It is estimated that insects destroy many millions of dollars worth of crops each year. It may safely be stated that the chief food of birds is harmful insects. On this account birds generally are very valuable to farmers economically as well as being a delight with their songs and calls. Birds deserve the protection and care of farmers. It is encouraging that we are beginning to realize the importance of birds, and that their wanton slaughter by boys, hunters and those seeking them to adorn ladies' hats, is being discouraged, and in many states laws have been passed to protect them.

Know Birds.—Every boy and girl should learn to know birds. A few birds are really harmful and should be destroyed, but only a few. Many suspected by casual observers of being harmful have, on close study, proved to be helpful. Even hawks and owls, which have too often been condemned, are generally worthy of protection. The only safe policy is either to know birds, so as to be sure that the ones killed are really harmful, or else assume that all birds are useful. Write to the U. S. Department of Agriculture, Washington, D. C., for Farmers' Bulletins Nos. 509, 513, 621 and 630. These bulletins will give much useful information about birds and how to attract and protect them. Write also to the Secretary of the National Association of Audubon Societies, at 1974 Broadway, New York, N. Y., for further information.

Attract birds to your homes by providing them with water, suitable houses, by supplying some of the winter birds with food during severe weather, by protecting them from cats, and by growing groves and shelter belts. There are few things more pleasant than to waken in the morning with the air literally filled with the music of birds. It is even more pleasant, if you know that some effort of yours has helped to make it possible for them to live and thrive in your neighborhood. It not only is pleasant, but it pays.

BEES

Importance of Bees.—The total honey crop of the country is not large; that is, it does not represent nearly so much wealth, as the wheat crop or dairy products, but nearly everyone likes honey, and honey is said to be one of the most wholesome sweets. While there are not a great many who derive large incomes or make a business of keeping bees, it is believed that more should keep them. Bee-keeping is light, pleasant work, and nearly every family in the country might at least have an abundant supply of fresh honey in return for a very little labor.

Varieties of Bees.—More persons would keep bees, if they were not afraid of being stung. The common black honey bee is often quite vicious and will sting at the least

Figure 115.—A well-developed apiary.

provocation. Italian bees are now quite common and, if handled at all reasonably, are not in the least vicious. Most bee-keepers handle the Italian bees with bare hands and bare faces and very rarely are stung.

Sources of Honey.—Honey is the nectar of flowers gathered by bees. White clover, sweet clover, buckwheat and basswood are the chief sources of honey, though the bees will gather honey from a great variety of flowers, if none of the choice plants are available. Honey is a product that, unless gathered and stored by bees, is entirely lost. Bees not only gather honey, a useful product, from a source that costs nothing, but they are of material benefit to many of the plants they visit, as they aid in the pollination of the flowers.

A hive of bees on every farm is a good slogan. To know about bees and their industrious, orderly habits is an inspiration. In every hive there is a queen, several drones, and several thousand workers. The workers go through a regular course of training that fits them for their work. Then they work industriously as long as they live, which is usually only a few weeks. Write to the U. S. Department of Agriculture at Washington, D. C., or to your State College of Agriculture for information about bees.

Questions:
1. Name as many of the common birds of your community as you can and tell whether they are useful or harmful.
2. How may one encourage the nesting of birds about one's home? Why is it advisable to do this?
3. Tell what you can about bees.

Arithmetic:
1. It has been estimated that insects cause a loss of $700,000,000 annually to farm crops in the United States. How much does this loss amount to per farm? (There are 7,000,000 farms in the United States.)
2. If, by destroying the birds, this loss from insects were doubled, what would be the total loss in the United States? What would be the total loss per farm?
3. If, by protecting useful birds and increasing their numbers the loss could be reduced 50%, what would be the total saving in the United States? What would be the total saving per farm?

CHAPTER XX

AGRICULTURAL ENGINEERING

THE ROAD PROBLEM

Importance.—Few boys and girls realize the great importance of roads. As we walk over them to school day after day we are likely to think how bad or how long they are, rather than to think what they are for, how they are made and how maintained.

Uses of Roads.—Roads are used as a means of communication, and as they become better it is easier for persons to travel and to haul loads upon them. Where roads are good, therefore, we can with less effort and much greater comfort go to school, to church, to town and to our neighbors. Such conditions make life pleasanter in the country and have a very strong tendency to make property more valuable. Good roads have an educational and social influence of as much importance as their economic value.

Figure 116.—A well-constructed turnpike.

Cost of Roads.—We often hear the remark that roads are poor, because it costs too much to build good ones. But did it ever occur to you that bad roads may cost more than good ones? All products of the farm must be transported over roads, and it may cost more to haul these products to market for several years, over poor roads, than to build good roads and haul the products over them.

Points to Consider.—Are the roads good between your home and the schoolhouse? Between your home and town?

Figure 117.—A specimen of a bad road.

Are there steep hills, or places where the road is rough, or muddy, or sandy? Have you thought that the size of the load that can be hauled to town is determined by the size of the load that can be hauled over the worst place or places? This is true; and many times one has to go to town with only half a load, on account of some bad place in the road. The cost of marketing farm products is thus increased and consequently the price of farm produce tends to increase; because, if the roads are bad, fewer products can be brought to town and fewer farmers will try to get products to market. Persons who live in town and have to buy the farm products are interested in good roads, because they want to get their vegetables, flour, etc., as cheap as possible. For this reason it is right to tax all the people in the county, state or in the United States, for the purpose of building good roads, because all the people are benefited by them.

Roads are nearly always four rods wide. This width consumes a great deal of land. A strip two rods wide is taken off all land adjoining a highway. Roads are made this width to give ample room for turning and to allow space for ditches, cuts and fills.

Questions:
1. In what way do poor roads affect the price of farm products?
2. Why is it right for those living in town to pay part of the expense of building good roads?
3. Explain how poor roads may be more expensive than good ones.
4. What are some of the advantages of good roads besides making it easier to market farm products?

Arithmetic:
1. If ten teams pass over a road each day, how many trips will be made over the road in a year?
2. If 3,650 trips are made over a road each year, and there would be a saving of 2c. per mile each trip, if the roads were good, how much would be saved per mile per year?
3. If $1,000 per mile were invested in good roads, and $73 were saved annually thereby, how long would it take to pay the $1,000, drawing 4% interest by applying the $73 saved annually?

ROAD CONSTRUCTION

The object in view in road building is to make the road bed as near level as possible, that is, to avoid hills; also to make and keep it as firm and unyielding as possible with the material and labor at hand. It is sometimes no farther around a hill than over it, and in such cases it is much more practical to go around. A pail handle is often used to illustrate this point. When standing erect the handle is the same length as when lying down, but a road, as represented by the erect handle, would be much harder to travel than a road represented by the handle lying flat.

Stone Roads.—In the older countries, and in the older and more thickly populated portions of this country, a large portion of the roads are built of some hard material, as stone. A very common form of stone road is called macadam. It is named after the man who invented this process of road building. To build a macadam road, the bed is first given the slope desired, then covered with a layer of coarse, crushed rock, which is rolled with a heavy roller. Then another layer of finer crushed rock is placed on top, and rolled until it works in between the particles of the coarser material. More, but still finer, crushed rock or sand is added, sprinkled with water and rolled until a smooth, hard surface is formed. A stone road made as previously described, and from 6 inches to 12 inches thick, makes an excellent, hard, permanent road. Such roads cost so much ($3,000 to $6,000 per mile) that they can be built only where the population is dense and where there is a great deal of travel over them.

Figure 118.—A split-log drag faced with steel. (See description and cost page 252).

Earth Roads.—In most farming districts, for many years to come, roads must be made of the material at hand; which means, in most cases, common earth. Such roads, if properly made and maintained, are very serviceable and may be much better than country roads generally are.

Drainage.—Since the object in making roads is to keep them hard, it is plain that, to do this, water must be kept from standing at or near the surface. Drainage, then, is the first problem in building roads (except sandy roads) and it is safe to say that, if all roads were properly drained, the greatest problem in road building would be solved.

The Turnpike.—The most common form of road is a turnpike, made by taking earth from each side of the road and putting it in the middle. This makes a very good form of road, as the center of the road is high, so that the water runs off to the sides into the ditches constructed there. Often water remains in these ditches, because no outlet is provided by which it can escape into the natural waterways. Water standing beside a road, and within two to four feet of the surface, is very often detrimental to the roadbed, as it soaks up through the earth and keeps it soft. It is sometimes necessary to ditch across some farmer's land to get the water out of these roadside ditches. Farmers should willingly co-operate with the town board to do such work. Instead, they sometimes object. But in most states the town board is authorized by law to construct such

Figure 119.—Cutting weeds along the roadside.

ditches as are necessary across any property. It, of course, must pay damages, if such ditch does damage the property; but, if it proves a benefit, then the owner of the property must help pay the cost of constructing it.

Where the road is made mostly of clay, it is greatly benefited by the addition of sand, as the mixture is less sticky, sheds water better and dries out more quickly. Likewise sandy roads are benefited by covering them with clay, as the clay helps to bind the sand together and keep the road hard.

If good gravel is at hand, that is sharp and will pack together, almost any of the common earth roads will be benefited by a coating of it.

Good gravel roads are better than earth roads, but not so good as stone roads. They are much cheaper than stone roads; and many communities are graveling a few miles of road each year, thereby gradually securing a very serviceable system of roads.

Questions:
1. What is the chief object in view in road building?
2. Tell how a macadam road is built.
3. What is the first problem to consider in constructing earth roads?
4. Why is a turnpike a good form of road?

Arithmetic:
1. If a road is 4 rods wide, how many square feet of surface are there on a mile of road?
2. If there are 30 in. of rainfall in a year, how many tons of water fall on a mile of road in a year?
(A cu. ft. of water weighs 62.42 lbs.)
3. How many cubic yards of gravel are required to cover a mile of road 12 ft. wide and 6 in. deep?

MAINTENANCE OF ROADS

Road Repairing.—In well settled communities the main part of road work is to keep roads in repair. Repair consists in fixing bridges and culverts, filling in ruts and mudholes, opening old and making new side ditches, and smoothing off and rounding up the roadbed so it will readily shed water. After an earth road is well made—that is, made as level as practicable and built into a turnpike with ditches on either side—there is nothing so cheap and effective for keeping it in repair as the split log drag.

King's Split-Log Drag.—Every one interested in good roads should know of the King split log drag. It is named after Mr. D. Ward King, of Maitland, Missouri, who first made known to the public the value of this excellent little implement. It is made of a log ten or twelve inches through and about eight feet long, split in halves. The halves are fastened together by boring two-inch holes through them and driving in strong stakes two and one half to three feet long, just as a wood-rack bed is made, with both split surfaces of the log facing the same way. If a log is not at hand, a timber about three inches by eight inches may be used instead. The efficiency of the drag is increased by putting a strip of steel on each cutting edge as shown in Figure 118. The drag then cuts better and wears much longer. It is drawn by a chain in the direction of its faces, and at an angle, so that it pushes earth toward the center of the road just as a reversible road grader.

Why Earth Roads Need Dragging.—When a road is first made it has a gradual and continuous slope from the center towards the sides. No place is left for water to stand on the road, and it soon dries off after a rain. As heavy loads are drawn over earth roads, the wagon wheels cut into the surface and throw up a ridge just outside of where the wheels run. You can see this on almost any road, especially after a rain. If these wheel tracks are allowed to remain, when it rains water will stand in them and soften the roadbed. Then as wagons pass over them they are made much deeper. The road drag is the simplest way of filling these ruts. It is cheaply constructed, and one man and two or three horses can manage it.

When to Use a Road Drag.—You have no doubt heard of "puddling" soil—working it when it is wet. Farmers sometimes make reservoirs for water in clay soil by excavating a hole, wetting the soil in the bottom and tamping it or leading horses or cattle about in the muddy bottom. This puddling makes the soil hold water. Since we want the surface of the road to be impervious to water, it is desirable to have it puddled. This can best be done by dragging it soon after a rain, when it is still wet. If a puddled surface will hold water, as in the case of reservoirs, it will also shed water when rounded and smoothed, as on a dragged road. There are also other reasons for dragging at such a time. The surface of the road is soft and the ridges are more easily cut off and pushed to the center. Men and teams can not work to good advantage in the fields, and the road dries more quickly. When all main traveled earth roads are dragged soon after every heavy rain, roads will be very much better than they are now, and the cost of this work is so slight that any well settled farming community can afford it, or each farmer can well afford to drag the road along his property.

If a hole is to be filled in a road, material similar to the road should be used, i. e., it is not wise to fill holes in a clay road with sand or holes in a sandy road with clay, as they do not wear uniformly and so make the road rough.

Sandy roads are best maintained by keeping them covered with straw or other vegetable matter, as this helps to hold moisture, and sandy roads are firmer when moist than when dry. The Minnesota State Highway Commission suggests that the road supervisors in sandy sections sow some strong growing crop in the right of way, to cut and throw in the road.

Questions:
1. Describe a King split-log drag.
2. For what reasons do earth roads need dragging?
3. What is accomplished by dragging a road when it is still a little wet?

Arithmetic:
1. How many feet of lumber in two timbers 3 in. thick, 10 in. wide and 8 ft. long? How much is it worth at $30 per thousand feet?
2. How much will a strip of mild steel, ⅜ in. thick, 2½ in. wide and 16 ft. long, weighing 3 lbs. per ft. cost at 3c. per pound?
3. A boy can make a road drag with the above materials in 5 hours. His time is worth 10c. per hour. What is the total cost of the drag?
4. A boy with 3 horses can drag a mile of road in 1 hour. How much will it cost, if the boy's time is worth 10c. per hour and each horse's time is worth 9c. per hour? How much will it cost, if he drags the mile of road five times? If he drags it ten times?

DRAINAGE

Drainage is the process of opening up a channel by which the surplus water in the soil may run off by the force of gravity. In hilly or rolling land drainage is provided naturally, as the water runs off over the surface. Sandy soils with sandy or gravelly subsoil seldom require artificial drainage, because the surplus water easily runs down through the ground. Flat land, or heavy clay land, or even sandy land with a heavy clay subsoil that the water cannot get through, often requires drainage. There are two general ways of draining land, surface drainage and tile drainage.

Drainage where needed affords one of the most profitable investments on the farm. Too much water fills the spaces between the soil particles and crowds out the air. We have learned that seeds and the roots of plants require air that they may grow. The roots of common field crops will not grow in soil that is filled with water; that is, where there is so much water that it crowds out the air.

Surface drainage is provided by digging ditches through and from a field needing drainage to a lower place called the outlet, as a ravine, a river, or a lake. Unless land is higher than the lake or river to be used for an outlet, it cannot be drained in the usual way. A ditch, to be effective in draining land, must start at the lowest place to be drained and continue in a downward slope to the outlet. Drainage water can run of its own accord only down hill. Surface drainage is the cheaper method and the more common. The chief objection to it is the continual work necessary to keep the ditches open. Grass and weeds grow in them and clog them up and earth washes into them or is worked into them by cultivation. Open ditches are also troublesome to work around. The space occupied by the ditch is wasted, and weeds are likely to make it unsightly.

Tile drainage is the more expensive form of drainage, but is very much to be preferred. Tile are pipes three or more inches in diameter, one foot long, and are made either of clay or concrete. Clay tile are at present more common than concrete. Tile are laid in the ground two and a half feet or more deep, and on a gradual grade or slope, so that any water that gets into them easily runs down the tile to the outlet. Good tile drainage work requires the use of rather accurate leveling instruments, and careful work in laying the tile to insure a continuous slope, so the water will all run out. Good tile, well laid, will last almost indefinitely. They are covered with soil, so that there are no waste land, irregular fields, or unsightly strips of weeds, as is the case with open ditches.

Questions:
1. What do you understand by the term drainage?
2. What are the advantages and disadvantages of open or surface drains? Of a tile drain?

Arithmetic:
1. If drain tile are 1 ft. long, how many rods of drain will 100 tile lay?
2. If tile cost $30 per 1,000, how much does enough tile to lay one rod cost?
3. If an acre of drained land yields 50 bus. of corn per year worth 50c. per bushel, and it costs $15 per acre to grow the crop, how much profit will there be per acre? How many similar crops will it take to pay for its drainage, if it cost $25?

IRRIGATION

Irrigation is the application of water to land artificially. In many parts of the United States there is not enough rainfall to ensure crops. In other parts there is practically no rainfall. In these places, if crops are to be grown, water must be supplied artificially.

Sources of water for irrigation are rivers, reservoirs, wells and lakes. Many rivers start up in the mountains and flow down through lower and flat country. For irrigation purposes, these rivers are dammed up, or a part of the water is diverted from the main stream and carried by means of ditches to the land to be irrigated. Reservoirs are often constructed in some position higher than the land to be irrigated. Water from snow or rain on the hills higher than the reservoir is caught in the reservoir. As it is wanted, it is conveyed in ditches out over the land to be irrigated. Water is sometimes pumped from wells, streams or lakes, and thus raised high enough so that it can be carried in ditches to fields needing irrigation. Water may be pumped directly into the ditches as needed, or it may be pumped at any time and stored in reservoirs for use later.

Distribution.—Watering a lawn with a hose or a flower bed with a sprinkling can is irrigating, but this means of applying water to the soil is naturally limited to very small areas. During a crop season it is necessary to apply water equal to several inches deep over the land to ensure a good crop. We have learned that it requires more than 500 barrels to cover an acre one inch deep. To make irrigation practical, some very easy way of distributing water is necessary. The most common way of distributing water is by means of open ditches. Large ditches with high banks are dug through the fields to be irrigated. By damming up the ditches the water can be raised in them a little higher than the surrounding fields. Then small lateral ditches a few feet apart may be opened at the sides of the big ditch, and the water will then flow out over the land and settle into the soil just as rain. In a few places water is carried over the fields in tile laid just under the surface of the soil. The water comes out of the joints in the tile, and is absorbed by the soil. This method is called sub-irrigation.

Uses of irrigation are to furnish moisture, and sometimes plant food for growing crops, and it is sometimes used to wash out of the soil undesirable salts (called alkali) occasionally found in large enough quantities in soils to injure plants. By applying large quantities of water, it dissolves these salts, and if the soil is tiled or has a porous subsoil, they are carried away and the soil will become productive.

Advantages of irrigation are a sure supply of moisture, which is not always the case when farmers depend on rainfall. Water may be applied just when it is needed, and in just the amounts needed. In such sections farmers are very seldom bothered by rain when haying or harvesting. The disadvantages are the cost of the water and ditches, the labor of applying the water, and the bother of the ditches in cultivating.

Questions:
1 What is irrigation? What experience have you had with irrigation?
2. What are some of the sources of water for irrigation?
3. Describe two or more ways by which irrigation water may be applied to the soil.

Arithmetic:
1. How many cubic yards of earth would be moved in digging a ditch 3 ft. deep and 3 ft. wide, 80 rods long? How much would it cost at 15c. per yard?
2. How many acres of land in a field 400 ft. square? How many feet of furrows or small lateral ditches would be needed to irrigate such a field, assuming there was a main ditch along one side and that the furrows were 15 ft. apart?

FARM MACHINERY

Machinery for doing farm work has been improved wonderfully in the past half century. In fact, there is no other part of the business of agriculture that has improved so much as the machinery used. It is only a very short time ago when plows were very crude, and crops were planted, harvested and threshed by hand. The people of the world owe much to the men who have given their time and energy to the improvement of farm machinery. The benefit has been equally valuable to people living in cities as well as to farmers. When farm crops were produced largely by hand labor, one man could handle only a very few acres of land. Now, with modern machinery, one man can handle, in the production of general farm crops, one hundred or more acres and still work no harder, and probably not so hard as formerly. The effect of improved farm machinery on agriculture is shown by the following.

The Bureau of Statistics of the United States Department of Agriculture published a statement to the effect

Figure 120.—Four good horses on a gang plow.

that in 1855 it took four hours and thirty-four minutes of human labor to produce one bushel of corn; while in 1894, a bushel of corn could be produced with but forty-one minutes of human labor.

Investment in Machinery.—At present there is a comparatively large investment in farm machinery on every farm, and quite a considerable part of the farm earnings must be expended for repairs for operating expense of machinery or for new machines. Farm machinery is, therefore, well worth studying. Sometimes machines are purchased when they should not be, and sometimes a needed machine is not purchased when it would be real economy to buy it. To determine whether or not to buy a machine requires a little careful study of the facts in the case. It is simply a question of what is the cheapest and best way of doing the work to be done.

Example: A farmer may have twenty acres of corn to cut each year. He has choice of several things to do. He may be able to hire it cut by hand, and can find out about how much it will cost him. He may be able to hire a neighbor who has a corn binder to cut it. He will know how much this will cost. He may be able to hire a binder by the day, or by the acre, and cut it himself. He can figure the approximate cost, or he may buy a binder and cut it himself. To determine the cost when he owns the binder himself, he must figure in interest on investment. (A new corn binder will cost about $125.) He must figure depreciation. (A corn binder depreciates about 10 per cent per year.) He must figure repairs, twine, oil and cost of shelter, also the labor required to cut the twenty acres of corn. In this way he may determine rather accurately which is the cheapest way. He must not overlook, however, the advantage and satisfaction that comes from owning a machine and having it to use just when he wants it without waiting for someone else or having to spend time looking for a machine or helping to do the work.

Using Machinery.—Some persons can get much more service out of machinery than others. To do good work a machine must be in good condition, well oiled, well sharpened, if it is a machine that cuts, the parts all adjusted so that there are no loose joints or bearings that do not run freely. One used to a machine can tell instantly by the sound or work of it whether it is running properly or not. A machine that is not in good condition is wearing out much more rapidly, and hauls much harder than one that is in good condition. A good machinist likes machinery, enjoys seeing it run well, and will repair it at once when it is out of order. Such men get good service out of machinery, do good work with it, and find real pleasure in running it.

Shelter.—One of the very common causes of loss and short life of farm machinery is lack of shelter. Some machines and parts of machines are not seriously injured by standing out of doors, while other machines deteriorate as rapidly standing idle out of doors as when in use. Machinery left out of doors not only depreciates in value, but it runs harder, is unsightly, is usually not where it is wanted, and parts are often found missing or broken just when the machine is wanted. Modern progressive agriculture that appeals to strong industrious men and women requires that convenient, serviceable shelter be provided for all farm machinery, and that each machine be in its place under cover when not in use.

Questions:
1. What can you say about the improvement in farm machinery?
2. Tell what you can about operating machinery.
3. Give as many reasons as you can why farm machinery should be sheltered

Arithmetic:
1. If it required in 1855 4 hours and 34 minutes of man labor to produce a bushel of corn, how many bushels could a man have produced in one day of 10 hours?
2. If a man can now produce 1 bushel of corn in 41 minutes of labor, how many bushels of corn can be produced in one day of 10 hours?
3. Find the annual cost of a corn binder, including depreciation 10%, interest 6%, and repairs 2%, the binder costing $125. What is the cost per acre, if one cuts 20 acres of corn per year?

FARM BUILDINGS

Importance.—Farm buildings represent from 10% to 30% of the value of all farm property. In other words, on a farm worth $10,000 the buildings are worth usually from $1,000 to $3,000. On the average farm from $100 to $300 is spent each year in building new buildings and repairing and remodeling old buildings. Buildings on the farm provide shelter for family, stock, machinery, crops for sale and for feed. There is probably no other thing on the farm that has so much to do with the appearance of the farm as the buildings. For the above reasons, namely on account of the cost, the use and the appearance of buildings, it is very important that very careful attention be given to the planning, arrangement, construction and maintenance of farm buildings.

Permanence and Cost.—In newly settled sections it is

usually necessary, on account of cost and uncertainty as to markets and types of farming, to build rather cheaply and temporarily. After a farming community is well established, the farms developed and equipped with live stock and machinery, and the farm is earning a fair income, it usually becomes advisable to plan buildings carefully, and build them of more permanent materials, so that repairs and depreciation will be lessened. A log building or a cheaply constructed frame building may be expected to

Figure 121.—An attractive barn.

last ten or fifteen years. A well built frame building, a brick, or a concrete building will last fifty or more years. The first cost of the more permanent buildings will be greater, but the annual cost will probably be no more. Because these better buildings are to be used longer, it is more important that they be well planned so that they will serve their purpose in the best possible way.

Planning.—A great many farm buildings are put up without careful planning. Farm buildings may be just as artistic and attractive as buildings anywhere, and, if so, will make the country more attractive and help to offset the pull to the city. Likewise farm buildings are used every day in the year in doing the work of the farm, and if so planned as to facilitate the work in the home and the work of caring for the stock, they will be of much greater real value. Architects have made a special study of planning buildings, and it is as a rule real economy to employ an experienced man to plan any buildings of importance. His knowledge of available materials, their strength and uses, and the proper arrangement of parts will often avoid expensive mistakes in building and usually give greater satisfaction and value for money expended than one can get without a carefully worked out plan.

Conveniences in the farm home, such as running water, heat and light are now becoming quite common in the older, better developed sections of the country. With the discovery of the septic tank that will handle and dispose of farm sewage with little cost and no danger to health, running water and plumbing may be had in country as well as city homes, where sewer systems are provided. (Your State University or the United States Department of Agriculture will furnish information regarding the construction of a septic tank.) The windmill or gas engine, now found on most farms, can pump water for the house as well as for stock. There is much work to be done in a country home, more on the average than in the city home, and modern conveniences that will lighten or make more pleasant this work should be as freely provided as modern up-to-date machinery is provided to lighten the work on the farm. Hot and cold water, so piped that it will run into a sink or bathtub when wanted and out again when one is through with it, is as great a convenience in the country as in the city, and may be provided just as easily. Electric or gas lights, gas for cooking, also hot air, hot water or steam heating plants are all available for use in the country, and as soon as finances permit should be provided.

Maintenance.—Unless kept in repair, buildings rapidly depreciate in usefulness, value, and appearance. It is well, if possible, to have a definite time each year to look over all the buildings on the farm and make any needed repairs. Paint adds greatly to the looks of buildings. It also protects them from deterioration. Buildings do not need painting every year, but usually should be painted once in three to five years. Window lights and loose hinges and boards should be repaired or replaced as soon as out of repair, likewise stalls, pens, floors and mangers should be looked after. Lack of attention to these things may cause serious loss or damage besides decreasing the value, utility and appearance of the property.

Questions:

1. What can you say about the importance of farm buildings?
2. Do you think the farm home may and should be provided with modern conveniences? Why?
3. What are some of the things to look after in keeping buildings in repair?

Arithmetic:

1. Determine as nearly as you can the total value of all the buildings on your home farm, or some farm with which you are familiar.
2. How many thousand shingles will it take to shingle a roof each side of which is 28 ft. by 80 ft., if 1,000 shingles will cover 125 sq. ft.?
3. If a gallon of paint will cover 50 sq. yds. of surface, how much paint will be required to paint a square house 30 ft. by 30 ft., 16 ft. high?

THE SILO

A silo is a receptacle with air-tight walls in which green, succulent feed, usually corn, may be put and kept in good condition until wanted. Silos are usually made round, because in this form they are stronger, and there are no corners in which it is difficult to pack silage. The material stored in a silo is called silage.

Importance.—A silo is recognized as an important part of the equipment of an up-to-date stock farm. It is no longer an experiment, as silos have now been in use more than twenty-five years in this country. If one will visit a number of farmers who have used silage, one will be convinced that the silo is practical on a fair-sized stock farm. Men who have used silos are their strongest advocates.

Advantages.—Silage is relished by nearly all kinds of farm animals. It is palatable, nutritious, and is a means by which the entire corn plant may be saved. With a silo one may frequently save an immature crop of corn that would otherwise be largely wasted. Mature corn, however, makes the best silage. The silo provides for the storage of a large amount of feed in a small space. It is a convenient means of storing feed, and silage may be fed at any time of year, summer or winter, or it may be kept over from year to year. So a good silo filled with corn silage insures one against a shortage of feed in winter or summer. Silage is often fed in summer to help out when pastures are short.

Kinds of Silos.—There are a great many kinds of silos. They may be built of concrete, of brick or lumber. There are agents in nearly every community who are advocating their particular kinds of silos. They quite often claim that other makes of silos than their own are not good. Many Experiment Stations, as well as farmers, have used all the common kinds of silos. When well made, they all have proved to be good. Good workmanship is necessary in a silo, because it must be air-tight, water-tight, have smooth walls, and be strong enough to withstand great pressure from the silage, and also be able to withstand the wind. Any silo that has these qualities will be found very satisfactory whether built of brick, wood or concrete.

Figure 122.—A concrete block silo.

Cost of Silos.—The most common size of silo is one that will hold about 100 tons of silage. Such a silo will usually vary in cost from $200 to $500, depending quite largely on the materials used and the prices of materials and labor in the community. If a farmer buys a silo from a company making a business of selling silos he will pay more for it than if he were to buy the materials and hire the building done, as is usually the case with other farm buildings. In other words, he can get a silo more cheaply by building it himself, because he saves the expense the silo company must stand in advertising and selling their silo; but for the extra money he pays for a patented silo, he is saved the bother of planning and is quite likely to get a good silo, because the silo company has usually built many silos and knows just how to do it. If a man is a good mechanic, or can hire a good mechanic who has had experience in building silos, he can usually save some money by buying his materials and building the silo himself; but, if a good mechanic is not available, it is much better and safer to buy some good patented silo.

Size of Silo.—Like other buildings, the silo must be planned to fit the needs of the farm. A cow will eat from 30 to 40 lbs. of silage per day. Other animals, like calves, sheep, steers, etc., will eat about the same amount in proportion to their weight. Ten 100-lb. sheep, one 1,000-lb. steer, or four 250-lb. calves will eat about the same amount of silage as a 1,000-lb. cow. One can tell about the number of animals on the farm and the number of days during the year they must be fed, and from that determine the number of tons of silage likely to be needed. There are plenty of books and bulletins that give the capacity of various sized silos, or one can figure out the approximate capacity quite easily by finding the number of cubic feet in the silo and multiplying this by 40, as a cubic foot of silage will weigh about 40 pounds. It is advisable to build a silo sufficiently large. A tall, narrow silo is more satisfactory than a low, wide one, because the deeper the silage the more it becomes packed on account of the pressure. The more silage is packed the better it will keep. The top of the silage in a silo is always exposed. If a silo is narrow there is less surface exposed.

Questions:
1. What is a silo?
2. What are some of the advantages of a silo?
3. Name as many different kinds of silos as you can.
4. What can you say about the proper size of silo to build?

Arithmetic:
1. If a silo costs $332.50 and holds 95 tons of silage, how much does it cost per ton capacity?
2. A man has 20 cows that will eat 35 lbs. of silage per day each. How many tons of silage will he need to feed these cows 200 days?
3. If an acre of corn produces 9 tons of silage, how many cows will it feed for 200 days, at the rate of 30 lbs. of silage per day?

FENCING

Kinds of Fences.—Fences of some description are found on nearly every farm. Sometimes these fences are in such poor condition that they are very little improvement to a farm, while on other farms they are straight, well built, well kept and a very great addition to the farm both in usefulness and in appearance.

Fences are used to keep stock either in or out of fields. Formerly fences were made of rails, but of late years timber is more scarce and other fencing material is being used. Barbed wire and woven wire are now comparatively cheap, easily put up, and so effective in enclosing stock that practically all fencing is of this material, even in timbered sections where rails are plentiful.

Figure 123.—A poorly-braced corner post from which it is impossible to stretch wires that will remain tight.

Fence Posts.—There is a great number of fence posts used every year, and, as timber becomes scarce, posts become more and more expensive. There are many different kinds of timber used for fence posts, and they vary in value according to their durability. Some kinds of posts will last from ten to twenty years before they rot, while other kinds will become useless in three or four years. As a rule, posts that last well are made of slow-growing timber, such as oak or cedar, while quick-growing timber, such as willow and cottonwood, rots very quickly when placed in the soil.

Posts deteriorate when set in the ground, by rotting. They usually rot off just below the surface of the ground, because here the soil keeps them moist and the air gets in from the surface, thus making conditions favorable for rotting. The top of the post does not rot, as it dries off too quickly, and the bottom of the post does not rot, because the soil keeps the air away from it.

Cement Posts.—Cement is now used for making posts, by mixing it with sand and water, then tamping it into molds of the proper shape. Such posts are very serviceable and get better the longer they stand. The only way in which they are injured is by breaking them. To prevent this, strips of wire are usually put inside of the posts as the mortar is being put into the molds. Cement posts are not in general use, as few persons have learned to make them. The first cost of cement posts is higher than for wooden posts. They are heavy to handle and some little difficulty

Figure 124.—A well-braced corner post that will always remain perpendicular and hold the wire tight.

is found in fastening the wire to them; but, considering their durability, they are not expensive and will probably be used to a great extent as their value becomes better known.

Steel posts are now used to some extent and will probably be used more than at present as timber becomes scarce and as more permanent fences are built.

A process has been discovered by which wooden posts may be treated with creosote and thus made to last two or three times as long as when untreated. This process is to dip the posts (or the part that is to go into the ground) in a vat of hot creosote. The creosote soaks into the wood and keeps them from rotting.

Investment in Fences.—Fencing is done entirely for live stock, hence the cost should be charged against them. Fencing intelligently done offers a good investment, but sometimes fences are built when they should not be. Unless there is sufficient live stock on the farm to require good fences, and unless the live stock are of such quality as to yield a profit from good feed, fences are not a profitable investment.

The larger the fields fenced, the smaller the amount of fencing required per acre, hence it costs less for fencing, if one has enough cattle to use a large pasture than when one has only a few head that can use only a small amount of pasture. Figure this out for yourself.

Investing money in fences is different from investing it in land, because fences deteriorate each year and after ten or twelve years must be replaced. If one invests $10 in land, it is probable the land will always be worth $10 or more, and the only cost each year is the interest on the investment. If one invests $10 in fences, the cost of the fence each year is interest on the investment and whatever depreciation there may be. If the fence lasts ten years, it is worth $1 less each year. Thus, to be a paying investment, the fence must earn about $1.60 per year to pay its cost, while the land must earn but 60c. per year.

Questions:

1. What is the chief use of fences?
2. Of what are they usually made now?
3. Upon what does the value of fence posts depend?
4. By what process are wooden posts made more durable?
5. Why must an investment in fences be regarded differently from an investment in land?

Arithmetic:

1. How many acres of land in a field 40 rods wide by 120 rods long? How many rods of fencing are required to enclose it? How many rods of fencing are required per acre?
2. How many acres of land in a field 80 rods square? How many rods of fencing are required to enclose it? How many rods of fencing are required per acre?
3. If fencing costs 25c. per rod and lasts ten years, what is the annual cost per rod? (Figure 6% interest on 25c. and add to it 1-10 of the cost of the fence.) How much is the annual cost per acre of such a fence, if 10⅔ rods are required to enclose an acre?

BUILDING FENCES

Good Workmanship.—In building fences, like most other kinds of work, a man can show whether or not he is a good workman. If one sees straight, well built and well kept fences on a farm, one expects, and is very likely to find, other things on that farm orderly and properly done. A fence often remains in place for many years. If it is crooked, it is an eyesore all those years. If it is straight and well kept, it is a constant source of satisfaction to the owner and to all who see it.

Corner Posts.—Barbed wire fences have been the cause of so much injury to animals that many farmers are strongly opposed to them. The greatest cause of injury to live stock is slack wires. These are not found where the wires have been properly stretched when the fence was built. In order to stretch wire tight it is necessary to have the corner posts set and braced firmly, so they cannot give and thus allow the wire to slacken. With corner posts set as in Figure 111, it is impossible to keep the wire tight; but, when they are set and braced as in Figure 112, they will always remain firm and keep the wire tight. Observe fences in your neighborhood and note those that have well braced corners.

Setting Posts.—Posts are usually set by digging holes with a post auger or digger, setting in the posts and tamping the earth firmly about them. It is especially important to tamp the earth very firm about the bottom of the post and just at the surface of the ground, as these are the two places on which the strain comes. Corner posts often need short pieces of plank spiked on them near the bottom, to keep the strain of the wire from pulling them out of the ground. See Figure 113.

Woven Wire.—Woven wire makes a much more desirable fence than barbed wire, as there is no danger of animals' being injured in it and it will serve for hogs and sheep as well as for cattle and horses. It is considerably more expensive than barbed wire, espeically if only cattle are to be enclosed.

Stretching Wire.—If corner posts are firmly set, it is comparatively easy to stretch either barbed or woven wire. The wire is fastened firmly to the post at one end of the line, then strung out and stretched. A great deal of time can be saved by arranging to reel out two wires at once. To do this, put two spools of wire side by side on a rod or crowbar in the rear end of the wagon, fasten the two wires and drive ahead, the same as when stretching one wire. Always stretch the top wire first, as you thus avoid tangling when the other wires are stretched.

There are several good wire stretchers on the market. If one has no other means handy, barbed wire can be well stretched by bracing a wagon, blocking up one hind wheel and winding the wire about the hub by turning the wheel by hand. A stretcher especially made for the purpose is necessary for stretching woven wire.

Cost of Fencing.—Any farmer should be able to tell approximately how much it costs to build any of the common fences on his farm. It is the annual cost per acre that is important. To find this, one must first take into consideration the number of rods of fencing required to enclose an acre. This, of course, varies with the size and shape of the fields. The cost of posts and wire is known, because they are usually purchased. If posts are cut on the farm, the cost of getting them out will represent their cost.

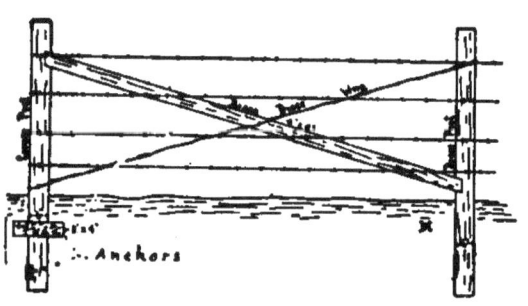

Figure 125.—A corner post braced one way and anchored so the strain of the wires will not pull it out.

Labor cost in setting posts and stretching wire can be found by experience or by asking others who have built fences under similar conditions. When one knows the total cost per rod for a fence, and the number of rods required to enclose an acre, one can tell the total cost per acre.

Questions:
1. For what reasons should corner posts be well braced?
2. Describe an easy method of stretching wire.
3. In what way may one determine the cost per rod of fencing?
4. If one knows the cost per rod, how can one tell the annual cost per rod? Per acre?

Arithmetic:
1. How many posts are required to build 80 rods of fence, posts 1½ rods apart? What are they worth at 12c. per post? What is the cost per rod for posts?
2. How many pounds of barbed wire are required to build 80 rods of 3-wire fence? (A single barbed wire 1 rod long weighs about

1 lb.) What is the cost at 3c. per pound? What is the cost of such a fence per rod for wire?

3. A man can set 50 posts in 10 hours. How long will it take him to set the posts in 80 rods of fence, if posts are 1½ rods apart? What will it cost, if his time is worth 18c. per hour? What is the cost per rod?

4. Two men and a team (2 horses) can string, stretch and staple 80 rods of fence (3 barbed wires) in 5 hours. What will it cost if a man's time is worth 18c. per hour and a horse's time is worth 9c. per hour? What will it cost per rod?

5. What, then, is the total cost per rod to build a fence as above? (Include posts, setting of posts, wire and stretching of wire, as found in above examples.)

CHAPTER XXI

COMMUNITY ACTIVITIES

BOYS' AND GIRLS' CLUBS

Importance.—Boys' and girls' clubs are becoming an important factor in the teaching of industrial subjects in both rural and city schools. A club tends to capitalize or at least make use of the combined interest and enthusiasm that comes from team play. It also makes use of one of the most important principles in education which is to find some means of arousing in the student a feeling that he needs or wants information that it is desired to impart to him. The average student can learn a thing that he feels that he needs at the time to know much more quickly than he can learn the same thing when he does not need the information at the time, and feels that it is more or less useless. The club undertakes to do something, and each member agrees to do his part. Each one is encouraged to go on with the work, because he knows others are doing it, and are expecting him to do it. Having before him a task that he wants to accomplish, he is receptive for anything that will help him to accomplish it. A bit of information or help just at the right time is appreciated and used.

Schools are the natural centers for boys' and girls' clubs and their activities. The teacher is the natural leader, and can be very helpful in organizing and directing the club. The club can be of immense help to the teacher. It helps to add to the school spirit and loyalty; it teaches citizenship and aids in discipline; it furnishes the needed stimulus to study; it helps to connect the school with the activities of the home, and shows the boys and girls that things learned in school are associated with and useful in everyday life.

Teachers are not all qualified, or feel that they are not, to teach some of the industrial subjects that are now quite generally required in schools; but the average rural school teacher is now qualified to teach elementary agriculture and need not hesitate to take up and study with boys and girls the rules and instructions now available in every state in the Union regarding the acre-yield corn contests, the bread-baking contest, the tomato-growing and canning contest, or the pig, calf or poultry contest. Reading over these instructions with her club members and helping them to understand them will maintain interest and give assistance

Figure 126.—A boys' and girls' farm club.

to the club members in their meetings and other activities. The instructions given in connection with the various club activities outline simply and clearly the best known farm practices for the community or state, practices that are usually better than the average for the community. The club provides a means for studying these instructions and for putting them into practice. When boys or girls have carried out the instruction given for the club project, they have accomplished something exceptional, and have learned important facts that will be useful to them as long as they live.

Time is limited in rural schools, but a good live boys' and girls' club tends to make better use of the time rather than to take valuable time needed for other classes. The club usually meets Friday afternoon once or twice a month. The time is put to good use, and it encourages so much work outside school hours that the time spent in study is increased rather than lessened. A great deal of the club work requires reading, composition writing and arithmetic, which subjects are better taught in connection with the club activities, because the pupils actually use them.

Figure 127.—A girls' sewing club.

Prizes.—Most of the club work has been based on prizes. There is a growing tendency to do away with prizes, or at least with large prizes. Most pupils will compete as strenuously for a club pin or a blue ribbon in a club project as they will for a "headmark" in spelling or a reward of merit for good behavior.

Object Acre.—Instead of large cash prizes, it has been suggested that the boys and girls in a community decide on something they would each like to do, such as to make a trip to the State Fair, take a sight-seeing trip to Niagara or Yellowstone, or take a term or course at some institution of learning. When the thing to be desired is determined upon by the club, then each member undertakes to produce a yield of sufficient value on a given piece of ground to pay his expenses. In this way each member has a chance to win. If he wins, he earns his own prize; if he loses, he can try again next year.

The value of boys' and girls' club work can hardly be overestimated. It is easily adapted to the needs or interests of a community, is easily put into operation, because boys and girls are naturally hopeful and enthusiastic and it very naturally works into the activities of the school. We feel that every teacher in grade or rural schools should give very careful thought to this work. A suggestive constitution and by-laws are submitted.

CONSTITUTION

Article I. Name

The name of this organization shall be........................ Boys' and Girls' Club.

Article II. Object

The object of this club shall be to improve ourselves, our school, our homes, and our community.

Article III. Membership

Any boy or girl in this district between the ages of 10 and 18 years may become a member of this club by signing the constitution.

Article IV. Officers

The officers shall consist of a president, vice-president, secretary, and treasurer, who shall perform the usual duties of such officers.

Article V. Meetings

The regular meetings shall be held at the schoolhouse the last Friday of each month during the school term, unless otherwise voted.

Article VI. Amendments

This constitution may be amended at any regular meeting by a two-thirds vote cast.

By-Laws

Section 1. The club motto shall be "To make the *Best Better.*"

Section 2. The officers of the club shall be elected by ballot at the first regular meeting of each school term, and shall hold office until their successors have been elected and qualified.

Section 3. The following order of business shall be followed at regular club meetings:

Roll call by secretary.
Reading of minutes of previous meeting.
Reports of committees.
Unfinished business.
New business.
Program.
Adjournment.

Questions:
1. What are some of the things a boys' and girls' club may do?
2. Why should boys' and girls' clubs be promoted by the school?
3. What do you think of the idea of the object acre? What could you grow that would enable you to take a trip to the State Fair?

Arithmetic:
1. If one produced 100 bus. of corn per acre and sold 25 bus. for seed at $3 per bushel, and the balance at 45c. per bushel, what would be the total value of the crop?
2. If one produced 300 bus. of onions on ½ acre, how much would they be worth at 60c. per bushel?
3. If one produced 200 bus. of potatoes per acre on 5 acres, and sold them at 50c. per bushel, how much money would one have?

FARMERS' CLUBS

A Farmers' Club is an organization of the people in any community for the improvement of themselves, their homes and their community. It should include in its membership the whole family, men, women and children. Two or more families may constitute a successful farmers' club; but it is best, where possible, to include all the people in the community. A rural school district is a suitable territory to be covered by a farmers' club. Meetings are held in the homes of the members, in town halls or in schoolhouses. There are many advantages in having the meetings at the homes of the members wherever it is practical to do so. The territory should be small enough so that all its members can easily convene.

Advantages.—A good, active farmers' club will do for a rural community just what a good, active commercial club will do for a village or city; namely, it will tend to secure the united influence of the community to bring about any desired improvement, and, further, it will unite the community to oppose anything that is not for its best interests. We can conceive of no way in which a farmers' club can be detrimental to a community, while we believe that there are at least four ways in which it may be helpful: (1) socially, (2) educationally, (3) inspirationally, and (4) financially.

Social Advantages.—People are essentially social beings. They are not usually happy when isolated, and do not develop properly except in groups. Life on the farm tends to keep people too much to themselves. A farmers' club that will bring the people together monthly or semi-monthly furnishes a very desirable change from the ordinary routine of farm life. Every one is interested in making the most of oneself and one's life. An important part of one's pleasure and development comes from meeting people and gaining the ability to mingle with them freely, without which one cannot appear at one's best or get the most out of life, either socially or in a business way.

One needs to get away from one's own work and home and get an opportunity to see things from a different angle. As a rule, a man is better satisfied with his own conditions when he sees how others live and do. A better acquaintance with people usually results in more tolerance for their shortcomings. Many times, when left to ourselves, we begin to think unkindly of our neighbors and really believe they are not what they should be. Usually a closer acquaintance and a clearer knowledge of their trials and struggles shows us that they are really better than we had thought them to be. A community in which all are interested in one another, know one another, and are boosting for one another and for the community, is a much better place in which to live than a community in which there is mutual distrust. As a rule, knowledge of one another increases confidence. Play is an important part of one's life. One cannot do one's best, if every minute is devoted to work. Relaxation and pleasure are absolutely essential to good living. Clubs that will bring some entertainment, social gatherings, or other means of amusement into the community, are very important.

Educational Advantages.—A good farmers' club may be of the greatest possible influence in broadening the knowledge of its members. The community has more information than any one of its members, and the club meeting tends to give each member the benefit of the knowledge and experience of every other member.

Being called upon to present various topics at club meetings stimulates study. No one farm or community has in it all that is good. Being forced to study what is being done in other places increases the general knowledge of the community and of each individual therein.

A farmers' club may increase the general knowledge of its members by bringing in outside talent. Business and professional men from the nearby towns or villages can be prevailed upon to address the club. Speakers from the University or the College of Agriculture and other public institutions may be secured occasionally to bring in outside ideas and inspiration.

Inspiration.—A discussion of the various problems of interest to the community always tends to stimulate every good, live citizen to desire better things, and to make a

Figure 128.—A co-operative laundry in connection with a co-operative creamery.

greater effort to secure them. Anyone who has confidence in people and in the community believes that almost all good things are possible, if the necessary effort and determination are put forth to secure them. If a club can succeed in arousing in its members a desire and determination for improvement in the community, better schools, better roads, better homes, better live stock, better farms, and better people are all possible.

Financial Advantages.—Business is now done in this country on a large scale. Millions of dollars and thousands of people are used in great enterprises. A farmer usually deals with people representing business interests larger than his own. As a rule, in business enterprises he deals with men who have the advantage, simply because the transaction means more to the farmer than to the other fellow with his wider field. For example, a potato buyer in a community may buy potatoes from two hundred farmers. What is 100 per cent of the farmer's business in potatoes represents one half of one per cent of the potato buyer's business. Consequently, a deal that means 100 per cent to the farmer means one half of one per cent to the potato buyer, and because the deal means very little to the buyer and very much to the farmer, the farmer is at a disadvantage. Exactly the same condition prevails in purchasing supplies. The farmer is handicapped on account of the small amount of business he is doing. A farmer who can use two dozen machines of one kind can purchase them more cheaply than the man who uses but one. The farmer who can sell many carloads of farm products of one class can get a better price for his products than the one who has only a wagonload or less to market.

If the products of a community, such as grain, potatoes and live stock, can be made uniform by co-operation among the members of the community in production, and then these larger quantities of uniform products can be sold by one man, the same advantages that come to the large farmer, or have come to the dairy industry, can be secured in other enterprises on the farm.

Co-operation.—A farmers' club is the logical forerunner of co-operation. In the first place, it gets the people of a community acquainted and increases the confidence of one in another. This mutual confidence is absolutely essential to successful co-operation. In the second place, it provides a logical means for studying carefully any enterprise that it is proposed to undertake co-operatively, so that impractical undertakings are likely to be avoided. We believe the farmers' club is a vital factor in promoting co-operation, because it is not organized to defeat any particular class of people, but to study intelligently any problem that may come up, and to execute it effectively.

SUGGESTED CONSTITUTION AND BY-LAWS

Article I. Name and Object

Section 1. The name of this association shall be the Farmers' Club of......................

Section 2. The object of this association shall be to improve its members, their farms, and their community.

Article II. Membership

Section 1. Anyone in good standing may become a member of this club by paying the annual fee of $........

Section 2. When the head of a family joins the club any member of his family may become an active member without paying additional fees.

Section 3. One third of the active members shall constitute a quorum for doing business at any regular meeting.

Article III. Officers

Section 1. The officers of this association shall consist of a president, a vice president, a secretary, and a treasurer. They shall be chosen for their business ability rather than their popularity.

Section 2. The officers of the club become the executive board and shall constitute the program committee.

Section 3. The executive board may call a special metting at any time by giving three days' written notice.

Section 4. The officers of this association shall be elected annually, and by ballot, at the regular annual business meeting, and shall hold office until their successors have been elected and qualified.

Article IV. Meetings

The club shall hold an annual meeting the................... Regular meetings of this club shall be held on the................of each month at the home of some member or at such place as shall be designated at a previous meeting, or by the executive board.

Article V. Amendments

This constitution may be amended at any regular meeting by a two thirds vote of the active members.

By-Laws

Section 1. The duties of each officer named in the constitution shall be such as usually pertain to his position.

Section 2. All other duties shall be performed by the executive and program committees.

Section 3. The club shall aid and further business associations among its members; particularly such associations as pertain to the purchase of necessary supplies, and the purchase and management of live stock and agricultural and garden products.

Section 4. From time to time the club shall give entertainments and hold meetings under direction of the program committee, for the benefit of its members and of those whom they may invite to attend.

Section 5. Any members, after due hearing, may be expelled from the club by a majority vote of active members at any meeting, without a refund of dues.

Section 6. These by-laws may be amended at any regular meeting by a majority vote of active members upon one month's written notice.

Questions:
1. What is a farmers' club? Name at least four ways in which a club may be helpful in a community.
2. How may a farmers' club benefit a community socially?
3. How may a farmers' club benefit a community educationally?

Arithmetic:
1. If there are 200 farms tributary to a town, each worth $12,000 what is their combined worth?
2. If 200 farms tributary to a town produce $1,500 worth of farm products, what is the total value of the products to be marketed?
3. If each of 200 farms purchase $1,000 of supplies, what is the total amount of their purchases?

CO-OPERATION

Meaning of Co-operation.—Every farm girl and boy should know the meaning of the term co-operation and realize its full significance. Co-operation means united effort or, as Prof. J. A. Vye has put it, "Union of the powers of the common people for the common good." The very best example we have of true and ideal co-operation is in the family. Here each member works for the common good, makes sacrifices for the rest, and shares in the joys and successes of the other members. In our business relations with neighbors and friends we cannot expect such complete co-operation. But, under present conditions of business, it is possible for people to co-operate or work together to their mutual advantage, even if they are prompted only by advantage to themselves and are without the generous desire to help others.

Obstacles to Co-operation.—One of the great obstacles to co-operation among farmers is the old notion that a farmer is the most independent man on earth. The farmer is independent in some ways. He is his own employer, may go to work an hour late or quit an hour early occasionally without asking anyone's permission; but he is dependent on others to buy his products; and, to get his supplies, he is dependent on manufacturers, transportation companies and merchants.

Need of Co-operation.—In the majority of cases the weaker of two persons making a trade comes out second best. The farmer, in selling his products and in buying his supplies, usually deals with large concerns; consequently he very often gets the worst of the bargain. This result is but natural, for the business of the average farmer amounts to but very little to the buyer of stock or grain, and, while the sale may mean a great deal to the farmer, it means very little to the buyer. Likewise, a dealer selling machinery, lumber or other supplies does not care a great deal about the business of one farmer, but the farmer is often under the necessity of buying of that particular dealer. In such cases it is quite evident that the farmer is at a disadvantage.

Advantages of Co-operation.—If, however, several farmers unite and offer for sale a large amount of grain, stock, or other products, there is business enough represented in handling this product to attract several buyers, and, as a consequence, a better price is obtained. Likewise, if a number of farmers find that they need among them several machines of different kinds (probably a few thousand dollar's worth) and they go to a dealer to buy, he is anxious to get the large order and will make some reduction in price in order to get it. Such combination of interests tends to equalize the strength on the two sides of the bargain, and, as a consequence, to equalize the benefits of the trade.

Co-operative Production.—Not only can farmers get better prices for the same product by selling in large quantities, but they can also offer, by working together, products that are worth much more. For example, if several farmers are raising horses, they can get better prices for them, if they all raise horses of the same type and breed. It is easier in such a case for a purchaser to get a matched team, and where several horses can be bought in one neighborhood, a buyer can afford to pay a better price for them than he could if he had to search several neighborhoods, because he is saved the time and expense of searching.

There are great possibilities for farmers to benefit themselves materially by closer co-operation in many of their farming and business enterprises. Several farmers working together greatly increase their own strength and resources, and better equip themselves to meet existing conditions.

They can put on the market large quantities of superior and uniform products in such condition and quantity as to assure top prices.

Questions:
1. What do you understand "co-operation" to mean?
2. Where is the best example of co-operation found?
3. What is a great obstacle to co-operation among farmers?
4. Why does a farmer trading with a large concern often get the worst of the bargain?
5. Why are a number of farmers, buying or selling together almost sure to get better prices?
6. Why can a buyer afford to pay more when large and uniform quantities of any one thing can be found in a neighborhood?

Arithmetic:
1. Thirty farmers want $100 worth of machinery each. How many dollars' worth of machinery do they all want? How much would they save, if they could get 10% discount? If they could get 20% discount?
2. A cattle buyer, to get a carload of cattle, spends 3 days, paying $5 per day for livery and $2 per day for hotel accommodations. His time is worth $5 per day. How much does it cost him in time and expense? How much does this expense amount to per head, if he buys 18 head?
3. If a car of uniform cattle, weighing 24,000 lbs., sell for 50c. per 100 lbs. more than mixed cattle, how much more are they worth than a car of mixed cattle?

MARKETING BUTTER

Co-operative Creamery.—One of the best examples of co-operation among farmers, in the Northwest, is the co-operative creamery; and the results should be sufficient to encourage effort along other lines.

Before the introduction of the co-operative creamery, every farm made its own butter. In most cases poor facilities were at hand for the work, and, as a consequence, a product often not first-class, and never uniform, was produced. The result was that butter sold at a very low price, often 6c. to 10c. per pound. Many farmers produced good butter, but very few farms were so situated as to be able to sell their product for a good price. Few farms had a constant or uniform supply of butter. It was impossible for a merchant to work up a trade for the product from a certain farm or of a certain quality, because he could not be sure of getting the butter for any definite length of time.

The result of this condition was that merchants in the small towns paid about the same price for all butter, mixed it and sold it at a low price. Consequently they could afford to pay but a low price for it.

Principles of Marketing.—There are three conditions which have a tendency to increase the price of a staple product, and, as these conditions are well illustrated by our creameries, we mention them here.

A large quantity of a desirable product in one place attracts several buyers, thus increasing competition and the

Figure 129.—A farmers' co-operative creamery where a large quantity of a uniform product is produced and offered for sale in an attractive and business-like manner.

price. A uniformly good product and a constant supply enables a dealer to build up a demand for something good; hence he can pay a better price for it. Bringing the buyer to the product, instead of sending the product to the buyer, tends to increase the price, because it enables the seller to refuse the price offered and wait for a better offer, which he can not always do, if he has delivered his product.

The average co-operative creamery produces from $20,000 to $50,000 worth of butter in a year. This amount naturally attracts buyers who are anxious to handle the large product. Creameries turn out a fairly uniform quality of butter, which qualification has been a very strong factor in increasing the demand and the price for creamery butter.

On account of the large and uniform product, buyers come or send to the creamery; consequently the creamery manager can sell or hold the product in the refrigerator until he gets the offer he regards as fair. These same principles hold true in the marketing of any farm product.

Why Some Creameries Fail.—That a well managed and well patronized co-operative creamery can compete successfully with any other known plant, in the manufacture of butter, has been amply proved. Still there are now many creameries that are being injured, and a few of them closed, by the competition of the large privately owned creameries called "centralizers." These "centralizers" have their cream shipped from any place they can get it. To get cream where there is a local creamery, they may offer a better price than the local creamery can pay, or they may get cream at the same price from a few of its dissatisfied patrons.

Losing any considerable amount of cream greatly weakens the local plant, and it cannot pay as much for butterfat as it had been paying, because the butter maker's salary and other expenses must be paid out of a smaller output. Thus the local plant is forced to run at a loss or to close.

Support Local Creamery.—The closing of the local creamery would not be so undesirable, if the centralizers continued to pay good prices for cream; but they cannot as a rule continue to pay as good prices as a well managed and well patronized co-operative plant, because they have difficulty in getting cream of as good quality as the co-operative creamery can obtain; and consequently cannot make as uniformly good butter.

It generally happens that when the local creamery is closed, the centralizer, being relieved of competition, reduces the price paid for cream below that paid by the best co-operative plants.

For their own interests as well as the interests of the community, it pays patrons of a co-operative creamery to stand by their own plant and not be lured away by temporary high prices, or high tests, or by jealousy and spite; for the chances are that, as soon as the local creamery is closed, they will get less for butter-fat than their own creamery can pay them, if they patronize it.

Questions:
1. What was the cause of the low price formerly paid for butter?
2. Give the three principles of successful marketing.
3. Why does a co-operative creamery sometimes fail?

Arithmetic:
1. If a farmer keeps 12 cows and each produces ½ lb. of butter-fat per day, how many pounds of butter-fat will he get per week? How many pounds of 25% cream will he get per week? (1 lb. of butter-fat will make 4 lbs. of 25% cream.)
2. How much should a farmer receive, if he sells 168 lbs. of 25% cream at 30c. per pound for butter-fat?
3. How much less would he receive per week for his 168 lbs. of cream, if he were paid but 27c. per pound for butter-fat? If he lost $1.26 per week, how much would he lose in 1 year?
4. How much less would he receive, if he were paid 30c. per pound, but his 168 lbs. of cream tested but 22% butter-fat?

MARKETING EGGS

Fresh Eggs Scarce in Cities.—It is no easy matter for people in town to secure good, fresh eggs whenever they want them. Thousands in every large city are willing to pay good, and even fancy prices for eggs, if they can be sure of getting a strictly first-class article.

How Eggs are Marketed.—The common way of handling eggs is about as follows: Eggs are gathered at irregular intervals, then about once a week they are taken to town and sold or traded to a grocer, who pays one price for all kinds of eggs—white, brown, small, large, dirty or clean—and mixes them all together. Some of these eggs are one day old and some are two weeks or more old. The merchant sets them in his storeroom with ill-smelling materials, as meats, oils, etc., and some of these odors are absorbed by the eggs. This mixed case, with other similar cases, is then sent to a commission merchant, who may sell them to a city grocer at once or store them. After more or less delay these eggs are offered for sale to city people, and it is little wonder that the housekeeper hesitates to buy them.

One Man's Experience.—A poultry man living near a large city sells all his eggs to a certain grocer. The eggs are gathered every day and the date stamped on each egg. Eggs of uniform size and color are put in small cartons or paper cases, holding one dozen each. The cartons are sealed with a label on which is printed a statement that the eggs are guaranteed to be strictly fresh, and if any bad eggs are found the producer will replace them with good ones. These eggs are marketed every day, and sell readily at 35c. per dozen, when eggs marketed in the ordinary way are selling at 18c to 20c. per dozen.

The Farmer's Problem.—Merchants who buy, handle and sell eggs are not to blame for the low price. It is the lack of uniformity, the unattractive appearance, and the suspicion that the eggs may not be fresh, that cause the low price. The farmer with a small flock of chickens can do very little to improve his markets alone, as he does not produce enough eggs to enable him to interest a grocer or to work up a special trade. This problem requires co-operation.

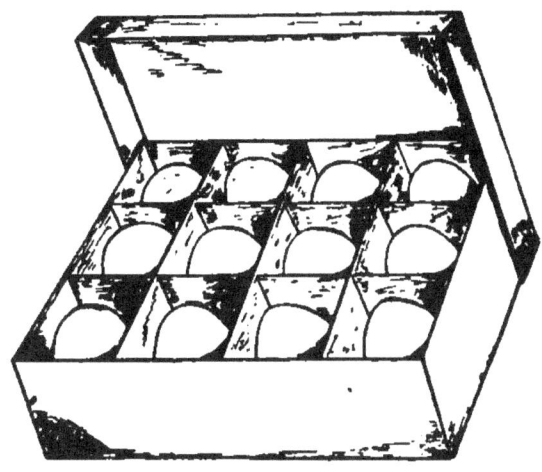

Figure 130.—A neat, attractive and convenient way to handle eggs.

The Barnum Plan.—At Barnum, Minn., eggs are marketed through the local creamery. Each patron having eggs to sell has a rubber stamp, with which he stamps on each egg the name of the Barnum Creamery and his own number, so that, in case the eggs are not good, they can be traced back to the right farm. Each patron delivers his eggs to the creamery with his cream. He must gather them every day, keep them in a cool, clean place, deliver no dirty eggs or eggs more than a week old. Thus the creamery company can guarantee the eggs to be fresh, clean and attractive. They have been shipping eggs in this way for years, direct to grocers in Duluth, who are glad to handle a high-class article. They have been able to pay their patrons from four to ten cents per dozen more than farmers get who market in the old way.

Suggestions.—A co-operative creamery is an excellent center at which to deliver eggs, where several farmers are interested.

The young people in a community can get valuable experience and remuneration by taking the matter up and marketing eggs as through the Barnum Creamery.

When a hen lays an egg in a clean nest, it is a clean egg. The next hen coming to the nest may soil the first egg. Gather eggs often.

THIS PACKAGE CONTAINS

One Dozen Guaranteed Fresh Eggs

BARNUM CREAMERY CO.

Manufacturers and Dealers

Eggs, Butter, Pasteurized Cream and Ice Cream

BARNUM, MINNESOTA

NOTE—Eggs in this package, if they have our trade mark on them, are guaranteed to be strictly fresh, clean and full size, and if ever found otherwise, we wish you would do us the favor to report it, giving number found on the egg.

BARNUM CREAMERY CO.

Figure 131.—The guarantee placed on each carton of eggs sold by the Barnum Creamery Co.

Grade eggs by size and color, and put in neat, clean cartons. Ship in good, strong cases direct to some good merchant in the nearest large city. Interest enough persons so that several cases can be shipped each time, and ship at least twice a week.

Try to interest producers in one good breed of chickens. Eggs will be uniform and more attractive.

Questions:

1. What can you say regarding the usual method of marketing eggs?
2. Why is the Barnum plan, suggested above, better than the old method?
3. In what way may the people of a community market their eggs to get better results than by selling them to the local grocer?

Arithmetic:

1. If each farm produces 30 eggs per day, how many farms would be required to produce enough eggs each day to fill a 30-dozen case?
2. How many farms, producing 30 eggs each per day, would be required to produce enough eggs, so that five cases, holding 30 doz. each, could be shipped three times a week?

3. If each farm produces 30 eggs per day, what would be the gain per day on each farm, if 5c. extra per doz. could be secured? If the extra trouble in keeping the eggs clean and neatly packing them takes 15 minutes each day, what rate per hour would a boy make for doing this work?

SCHOOL GARDENS

The school garden is a plat of ground in connection with the school, used by teachers and pupils for growing crops as a part of their school work. There are many who are enthusiastic about the school garden, and others who believe it has only a limited place as a means of teaching agriculture. A few years ago much was done to encourage rural school teachers and principals of consolidated schools to maintain gardens in connection with their schools. Many earnest attempts were made to carry out the suggestion, but at present there are very few places where the work is continued.

The advantages of the school garden are that it furnishes actual practice in the various operations necessary in the growing of crops. It tends to call the pupil's attention to many interesting and valuable facts about soils, plants, cultivation, also to insects and diseases affecting plants. It gives a practical trend to the agriculture taught in the school, and provides a means of teaching by doing.

The disadvantages are that the conditions of cultivating the soil and handling the crops are entirely unnatural. It is difficult to get the soil in good condition on account of lack of manure, teams, implements, etc. Usually about June 1st when crops and weeds are growing rapidly, school is out and no one is left to care for the garden. By fall the garden is likely to be a patch of weeds, and to illustrate something not desirable rather than something valuable.

Place for School Gardens.—In medium-sized and large towns, where there are many pupils from homes that do not or can not have gardens, and where a special agricultural instructor or gardener can be employed throughout the crop-growing season, there is a real need for school gardens. Under such conditions the school garden furnishes concrete illustration and practice in soil culture and the growing of crops that could be gotten in no other way. It also furnishes healthful, useful employment for a few hours each week to boys and girls who would otherwise be running the streets. With a capable instructor in charge, such a school garden may be very valuable as a field laboratory in connection with the regular classroom work, and may also be used to good advantage to furnish much material to be used in the classroom during the winter.

Where Not Needed.—School gardens have very little place in connection with rural schools, consolidated schools, or village schools where all of the pupils have gardens at home and some useful employment during vacation. Under such conditions the teacher or agricultural instructor can do more good by encouraging home gardens and by aiding the pupils in their garden work at home, than by attempting to handle a school garden.

Arithmetic:
1. What is a school garden?
2. Name some of its advantages and disadvantages.
3. Under what conditions are school gardens desirable? Where undesirable?

Arithmetic:
1. How much does it cost to plow ¼ acre of land, if it takes a man and 2 horses 1¼ hours to do it, assuming that a man's time is worth 20c. per hour and one horse's time is worth 10c. per hour?
2. How much does it cost to spade up ¼ acre of land, if it takes a man 20 hours to do it, a man's time being worth 20c. per hour?
3. If one can raise 600 bus. of onions per acre, how much can one raise on 1 square rod? (There are 160 sq. rds. in an acre.) How much would the onions be worth at 60c. per bushel?

COUNTY AGENTS.

A New Movement.—During the past few years a really new movement has been started known as the county agent movement. This is a valuable form of community work. It was started in the Southern part of the United States when the cotton boll weevil first made its appearance and threatened to destroy the whole cotton industry. The cotton planters appealed to the United States Department of Agriculture for assistance which was readily given. Men were sent into several communities, and gave the best assistance available in combating the pest. Their efforts were fruitful, so much so that other communities asked for the same help.

Funds.—The United States Department had not funds enough to enable them to extend the county agent work as rapidly as it was sought. Business men were quick to see the advantage to themselves and to the community in saving the cotton crop and readily donated money to supplement the United States Department funds. The work was rapidly extended and soon took on the form of county work, usually each county employing its own agent in cooperation with the United States Department of Agriculture and its State College of Agriculture. This work spread rapidly over the South and later was started in the North, so that in 1915 there were about 1,500 county agents employed in the United States.

The work of a county agent is varied. He is the joint representative of the people in a county, the State College of Agriculture and the United States Department of Agriculture. It is the agent's business to do everything he can for the improvement of agriculture, business and conditions of living in the county in which he is employed. He brings directly to the service of the farmers and others in the county the facilities of the two institutions they have created and are maintaining, namely, their own state Experiment Station and the United States Department of Agriculture.

Local Information.—As an agent travels over a county he finds many men who are succeeding in their particular lines of work. One man has made a success of corn growing. Another has succeeded in fruit raising. Others have succeeded with dairying, hogs, poultry or with beef cattle. The county agent with his thorough training and experience is able to study the methods of these successful men and to quickly determine the principles which have resulted in their successes. He is then in position to inform others of these successful methods, to suggest improvements in the general methods of farming in the county and in many cases induce farmers to visit the farms where the most successful methods are being practiced.

The county agent must be a man thoroughly familiar with farm work and farm life from having lived and worked on a farm for years. He must be quite familiar with the work of the Colleges of Agriculture and of the United States Department of Agriculture. He must be a man who can meet people easily and win their confidence and he must also be a natural leader and organizer. Such men are rare; but, as demand for county agents increases, men will be developed and trained for that particular work.

A Community Movement.—The success of the county agent movement in any particular county depends a great deal on the ability of the man selected as agent, but it depends a great deal more on the people in the community who are back of the work. The work is bigger than any agent. It represents a desire and determination on the part of the people in a county to improve and an organized effort to accomplish their desire. Success in farming is becoming more and more dependent on the success of the whole community. Good roads, good schools, good markets, good social conditions and in fact good facilities for the improvement of live stock and other farm products are beyond the power of individuals working alone and can be had only by groups of people working together in an organized way.

Future.—If one can judge of the future of the county agent movement by the results in the United States during the past three years, it is destined to be one of the most important and democratic system of Agricultural education yet attempted.

Questions:
1. What can you say of the county agricultural agent movement in your state?
2. Why is it important that the citizens of a county be organized before employing a county agricultural agent?
3. In what ways may a county agent be of help to the farmers of a county?

Arithmetic:
1. If 100,000 acres of corn are grown in a county, how much is it worth to the county to have seed corn that will yield 1 bushel more per acre than the ordinary seed commonly used, if corn is worth 45 cents per bushel?
2. If by proper vaccination a county agent may save 2,000 hogs in a county from loss by hog cholera. how much is saved, if each hog is worth $12?

CHAPTER XXII

THE FARM HOME

WHAT A DESIRABLE HOME SHOULD BE

Pleasant Surroundings.—Almost every one admires order, neatness and beauty in preference to their opposites—disorder, untidiness and barreness or lack of beauty. A farm seems more valuable, if the yards surrounding the house and barn are clean, neat and orderly. If one sees a smooth, well kept lawn, with a few appropriate shade trees and flowers, one feels sure that here is prosperity, or here, at least, is a restful home.

In cities and villages considerable attention is paid to beautifying the lawns, and a very little effort adds an elegance and charm to an otherwise plain home.

Facilities.—The farm folk may not have much time to think how they would enjoy these things or to establish and keep such conditions, but there are many things that can be had with but little effort and expense, and many which the boys and girls can do. In the country, where there is plenty of room for a lawn, for flowers, ornamental shrubs and trees, and where good soil, fertilizer, team and machinery for keeping up a lawn are available, it is to be regretted that these possibilities are so little realized.

Neatness.—The first step for the boys and girls, who are anxious to aid in beautifying their homes, is to keep things as neat as possible. Every boy of school age can move a pile of ashes, if the ashes during the winter have been piled too near the house. Raking the chip yard is not too difficult for the average boy. Often broken dishes, tin cans, old shoes, etc., have been cast into a pile somewhere. These should be removed and buried. Frequently the banking about the house is left until late into the summer, or only partly removed on account of frost when the work was begun. Persuasion from boys or girls might bring the team to the house at the close of a day's work, to remove the banking or the remaining portion of it.

The Lawns.—When neatness has been established, consider next the lawn. Many farmers have reasonably smooth lawns covered with good sod. If this is well burned over and thoroughly raked in the spring, it will not be difficult to keep it cut with a lawn mower all summer. The grass must be kept low or it will be difficult or perhaps impossible to cut it with a lawn mower. A lawn mower may sound

Figure 132.—View of a pleasant farm home, showing good effects of lawn, vines and trees.

like an extravagant article, but it is not expensive, considering the number of years one may be run.

Many farm yards are bare and packed hard from frequent travel. In such cases the aid of the father or "big brother" must be solicited to plow them up and perhaps grade them a little. If the soil is not rich enough, a few loads of manure and black earth might have to be hauled. The boys and girls can finish the work by thoroughly raking the soil to prepare it for the seed. Blue grass alone is the most desirable lawn grass; but, as this starts very slowly, it is well to mix with it some more rapid growing varieties, as white clover and redtop. Sow this mixture very thick, and rake again to cover the seed. Roll the lawn to make a smooth surface and give the seeds a better chance to start quickly.

The lawn should be plowed and seeded as early in time for the spring rains to start the grass. The farm lawns may suffer during a dry summer, but under ordinary conditions they will do well, if the soil is kept rich by an occasional dressing of well rotted manure.

Desirable paths may be made of gravel or sand.

A few good shade trees and ornamental shrubs add much to the appearance of a farm yard, and there is no reason why Arbor Day cannot be observed on the farms and many desirable trees set out. Soft maple makes a very beautiful shade tree and is of reasonably quick growth. The elm is a popular tree for the yard or driveway. The box elder is commonly used and makes very rapid growth.

Modern conveniences in the home are now as easily available in the country as in the city. The septic tank provides just as satisfactory means for disposing of sewage as a sewer system. A septic tank may be built for a few dollars on any farm. Gasoline engines for pumping and pressure tanks or other systems of water supply are no more expensive in the country than in the city. These things make it entirely practical to have modern homes in the country and wherever such conveniences can be afforded they should be installed. They save work in the house just as much as improved machinery saves it in the fields. A modern home in which there is running water, a good heating plant and electric or gas lights will do much to make the farm home attractive to both young and old.

Questions:

1. Give as many reasons as you can why a neat, orderly farm home is preferable.
2. What is the first step in improving the appearance of a farm home?
3. Tell what you can about starting a lawn.
4. Write what you can about shade trees.

Arithmetic:

1. How many cubic yards of black earth are required to cover a lawn 100 ft. square 1 in. thick?
2. If it takes an hour to mow a lawn, 100 ft. square, and it must be mowed 18 times during the summer, how many hours of time are required? How much is this time worth at 15c. per hour?
3. How far would one have to walk to mow a lawn 100 feet square, with a lawn mower that cuts a swath 16 inches wide?

WINDBREAKS

Value.—A good windbreak about a farm home is very valuable and is an inexpensive luxury. If there is a good windbreak about the buildings, less fuel is required to keep the house warm and the stable will be much warmer. Animals must be kept warm during the winter either by shelter or by feed. If they are exposed to the cold winds or are kept in cold stables, they must have more feed. Feed is too expensive to be given merely to produce warmth. The shelter belt really saves feed, which is worth money.

The windbreak also protects the orchard and garden from early frost, from storms, and from hot winds, making them much more likely to produce good crops.

It is a great comfort to hay-makers and harvesters to get behind a good windbreak, on windy days, to unload hay or grain. Many times it is possible to finish staking behind such shelter, when it would be impossible to handle the hay or grain out in the open.

The windbreak makes the task of doing chores much pleasanter and easier during the winter than it would be if the buildings and yards were exposed.

Kinds of Trees to Use.—There are many kinds of trees that make good windbreaks. The first requirement is that they be hardy—that is, will not kill out during the winter or during a dry summer; second, that they have a neat appearance and grow in such form that they really check the wind.

In starting a new windbreak, trees that grow rapidly are, as a rule, used. The white and the golden willow, cottonwood, Norway poplar, box elder, soft maple, etc., are some of the rapid-growing trees. These trees, if properly set out and cared for, will make a good windbreak in five to ten years. Trees that grow rapidly usually do not last very long; so, if the quick-growing trees are used, slow-growing and longer-lived trees should be set among them to replace them when they begin to die and break down. Some of the common slow-growing, long-lived trees are the elm, hard maple, green ash, and several of the evergreens.

Planning for the Windbreak.—In planning for a windbreak it is well to make a sketch of the farm, showing the buildings, lanes, fields drives, yards, etc., and so arrange the trees in setting them out as to give protection from the north and the west winds. If possible, arrange the trees so that the road and other points of interest may be seen from the house.

The windbreak should be far enough from the buildings

Figure 133.—An evergreen windbreak.

that snow will not drift about them or into the yards. It is well to have it include the garden and orchard as well as the buildings.

Questions:
1. Of what financial value is a windbreak?
2. What kinds of trees are best for windbreaks?
3. For what reasons would you have the shelter belt or windbreak a considerable distance from the buildings?

Arithmetic:
1. What is the cost of 1,500 willow cuttings at 75c. per thousand?
2. What is the cost of 460 spruce seedlings at $5.00 per hundred?
3. How many trees will be required to set three rows 20 rods long—trees 4 ft. apart in the row?
4. How many willow cuttings will be required to set 3 rows 60 rods long, cuttings 2 ft. apart in the row?

SANITATION

Healthfulness of the Home.—The first thing to be considered about a home is its healthfulness. No matter how beautiful, attractive and comfortable a house may be, if it is not a healthful place, no one would desire to live in it. One who is building a new house may select a site which will give drainage, ventilation and plenty of sunshine. One

Figure 134.—A farm home so situated as to afford good drainage. Shade trees are an excellent addition to a home, but should not be so close or so thick as to prevent a good circulation of air and the entrance of sunshine to the rooms.

who is buying a farm, expecting to live in the house already upon it, would do well to consider the healthfulness of the situation as well as the fertility of the soil. Many who find themselves in unhealthful homes, may often, with but little labor and expense, make the place quite as healthful as any they might have selected.

Drainage about the Home.—The healthfulness of a farm home depends first of all upon drainage. The ground should slope from the house, so that the cellar and yard may be dry. The well should be so situated that no surface water, or seepage from house, barn or any other building, can possibly get into it. It is of the greatest importance to the health of all in the home that the drinking water be kept pure. If the natural drainage is not sufficient, a system of drainage that would give the desired result should be carefully planned and put in. An open ditch could be made, but a tile drain is preferable. If it is worth while to drain land to produce better crops, it is worth while to drain to make the home more healthful.

Chickens, turkeys or fowls of any kind should not be allowed about the well or in the yard immediately surrounding the house. They are not only one more means of carrying dirt to the house, but they also destroy grass.

House flies are a great menace to health as well as to comfort. If there is disease in any home in the neighborhood, flies may carry the germs on their legs and bodies and so infect other homes. It is generally believed that they breed to a great extent in horse manure. For this reason, if for no other, manure should not be allowed to accumulate in the yards or about the barn.

Houses should be well screened against flies. Removable wire screens are the most desirable. While they may be a little expensive at first, they last so long that their yearly cost is very slight. If taken down in the fall and carefully put away for the winter, and then given a coat of thin paint before they are put up in the spring, they will last almost indefinitely. Where such screens cannot be had, mosquito netting may be used.

Bacteria.—Even when the outside conditions are all that they should be, a crusade against dust and germs must be kept up within. A few years ago germs and microbes were practically unheard of. To-day we hear and read much about them, and are just beginning to realize their influence. There are a number of terms in common use to-day which it may be well to understand. Bacteria are very simple, minute organisms belonging to the vegetable kingdom. They live in soil and water and on the skin of man and beasts. There are hundreds of species of them. Some species are helpful—such as cause decay of vegetable matter in the soil and so enrich the soil—other species produce disease. Many organisms, so small that they can be seen only with a microscope, whether animal or vegetable, are called germs, microbes and micro-organisms.

It is now known that these germs or microbes cause many of our contagious diseases, such as tuberculosis (consumption), diphtheria, and typhoid fever. The mistress of a household must remember that she cleans her house, not only that it may look well-kept, but more especially to make it sanitary. While dust and dirt in themselves may not be directly harmful, they are likely to become hotbeds for disease germs.

Questions:
1. What may often be done to improve unsanitary conditions, if they exist?
2. Why should not fowls be allowed about the house?
3. What are the objections to house flies, and how may they be kept out?
4. Explain the terms bacteria and germs or microbes.

Arithmetic:
1. How many square feet of wire netting are required to make a screen for a window 30 in. by 60 in.? What would the screen cost at 2c. per square foot?
2. What is the cost of screen at 2c. per square foot for a door 3 ft. by 7 ft.?
3. Screen window frames are usually made of lumber 1 in. thick and 2½ in. wide. How much lumber is required to make a full size screen frame for a window 30 in. by 60 in.? What is the lumber worth at $35.00 per thousand feet?
4. Screen door frames are usually made of lumber 1¼ in. thick and 3 in. wide. How much lumber is required to make a screen door frame 3 ft. by 7 ft.? What is the cost of the lumber at $35.00 per thousand feet?
5. What is the cost of material to make screens for a house having 3 doors and 14 windows? (Use sizes and prices given above.)

VENTILATION

Ventilation is Simple.—People and animals need pure and fresh air in order that they may be healthy. Supplying this fresh air is much simpler, especially in the country, than in generally supposed. It is known that air moves about easily, and that it presses down, due to its weight, upon the earth. If, for any reason, the pressure is not the same on all sides, the air will move in the direction of the least pressure. Every pupil knows that when a fire is lighted in a stove the heated air and smoke will rush up the chimney. When air is heated, it expands; hence it becomes lighter. This principle may be illustrated by heating a bottle so that the air in it will be quite hot. Then place your thumb or the palm of your hand over the top of the bottle and allow the bottle to cool. Cooling may be hastened by putting the bottle in cold water, keeping your thumb or hand firmly over the top. As the bottle and the air it contains cool, you will feel a pressure on the back of your thumb or hand, which will seem to be sucked into the bottle. This fact indicates that, as the air in the bottle

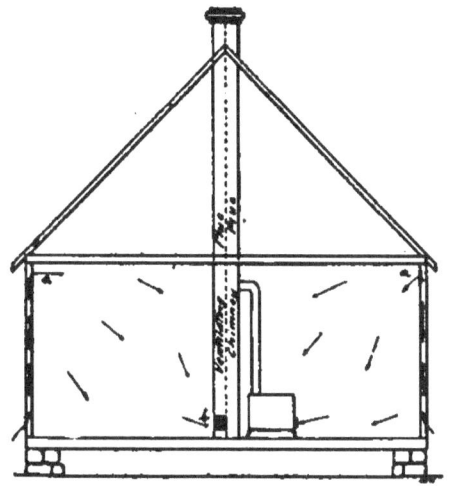

Figure 135.—Method of ventilating a schoolroom when fresh air is taken in through the wall near the ceiling. Air movement indicated by arrows.

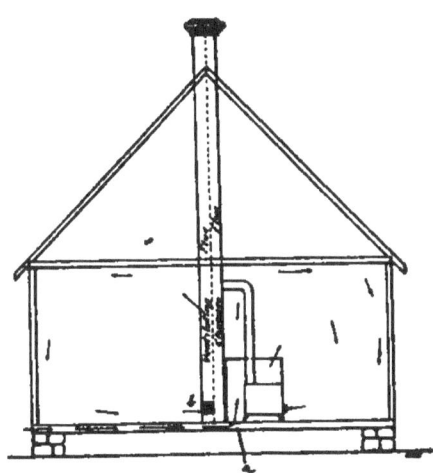

Figure 136.—Method of ventilating a schoolroom when fresh air is taken in under the stove, where it is heated, then diffused through the room. Arrows indicate movement of air.

cools, it contracts and occupies less space. It is evident, then, that cool air occupies less space than warm air.

How Air Moves.—Because air expands when it becomes warm, it is lighter in proportion to its bulk. The air moves somewhat as a pair of balances, the heavier end going down and the lighter end going up. When some of the air is heated and some remains cool, it gets out of balance, and the heavier cool air pressing down around the lighter warm air forces the warm air upward. Hence, what we call draught in the chimney is caused by the heavier cool air outside forcing, by its greater weight, the light, warm air upward.

Object of Ventilation.—The problem in ventilation is not to remove the warm air from within buildings and let in cool air, but to remove foul air and replace it with fresh air without causing a draught. Thus the principles explained above must be known and also the fact that foul air is heavier than pure air and is usually found near the floor of the room.

Ventilation Flue.—A stove with dampers open is very good for taking foul air from a room, as it takes air from near the floor. To provide ventilation when stoves are not in use, a double flue chimney should be built, with one flue opening near the floor to remove the foul air, and the other flue to carry the smoke from the stove or furnace. The heat in the smoke flue warms the ventilating flue and aids in the circulation of air. In this way foul air is removed from near the floor, while warm air, which naturally rises near the ceiling, is not removed until it becomes foul and settles to the floor.

Fresh Air Supply.—To supply fresh air is very simple in rooms heated by the furnace, as the fresh air is brought in from outside and heated in passing through the furnace into the rooms. In buildings heated with stoves, steam or hot water, air must be brought directly to the room from outside, and the object is to do this without causing a draught of cool air on the occupants of the room. There are two good methods of letting fresh air into rooms. The first and simpler is to let it in near the ceiling, as shown in Figure 126, where it will spread out over the room and gradually settle through the warm air, and as it becomes foul be drawn out through the stove or ventilating flue from near the floor. Another method, one frequently used in schools, is to have a sheet iron jacket about the stove. Air is let in under the stove, is warmed and rises between the stove and jacket to near the ceiling, where it spreads out over the room and is drawn from near the floor, as in the other case. See Figure 136.

Questions:
1. How can you prove that air expands when heated, and is lighter than cool air?
2. What causes air to move up a chimney?
3. What is the object of ventilators?
4. Describe two good methods of getting fresh air into a room.

Arithmetic:
1. A man needs 20 cu. ft. of air per hour, how much does he need in 8 hours? How much air will two persons need while sleeping in a room 8 hours?

2. If one needs 20 cu. ft. of air in an hour, how much will 40 children need in 6 hours?

3. How many cubic feet of air in a room 10 x 10 and 8 ft. high?

THE FARMSTEAD

Arrangement of buildings, windbreaks, shelter belts, orchard, garden, yards and lanes is worthy of close study. The sketch of a farmstead submitted herewith, Figure 128, may be used as a suggestion in working out a plan for any farm, though there are probably only a few places where it would entirely fit. This plan represents a great deal of thought and study and will serve as a guide in planning a farmstead anywhere.

Location.—Other things being equal, it is advisable to locate the farmstead along one side of the farm, rather than at one corner, because it gets the center of operations nearer to all the fields. It is also advantageous to locate buildings where there is good drainage; also near a main road and near neighbors, and where a good pleasant view is afforded.

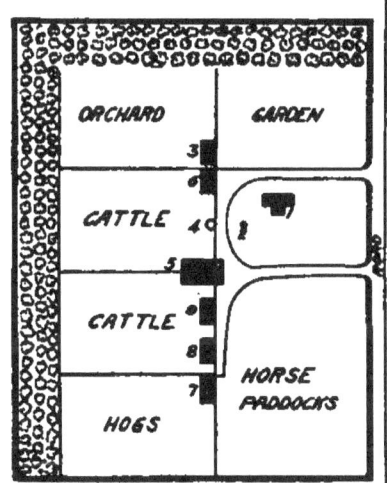

Figure 137.—Eight-acre farmstead facing east. 1 house, 2 well, 3 poultry house, 4 watering trough, 5 main barn, 6 machine shed, 7 hog house, 8 corncrib, 9 granary.

Size.—Many farmsteads are entirely too small. Six acres or more are needed for best results, so that the buildings may be far enough from the windbreak so that the snow will not drift about them. In the plan submitted, eight acres are used. Of this space nearly two acres are devoted to a windbreak. The house is located on a lot in front. This lot is made comparatively small, because there is not time enough to care for a large lawn. At one side of the house and next to the road a horse or calf paddock is provided. This is an open grassy patch that gives breadth and a broad view to the place, and makes it distinctly different from a city home on a crowded lot. Alfalfa or any hay crop may be grown on this plat, and cut for hay, if one does not like the paddock as planned. On the other side of the house is the garden. It is near the house where the women can easily get to it for fresh vegetables and small fruit.

Convenience.—The orchard and poultry house are just back of the garden, also to accommodate the housekeeper. The hog house and yard are farthest away from the house, because these are most likely to be objectionable. The corncrib is near the hog house, because most of the corn is likely to be fed to the hogs. It is also near one of the cattle yards, convenient to feed to the cattle, if desired. Next to this is the granary, which is also near the main barn, so that grain for feeding hogs or the stock in the barn may be near where it will be used. The main barn is next and directly in front of one of the drives to the main highway. This gives the barn a prominent place, and makes it convenient in driving to or from the barn.

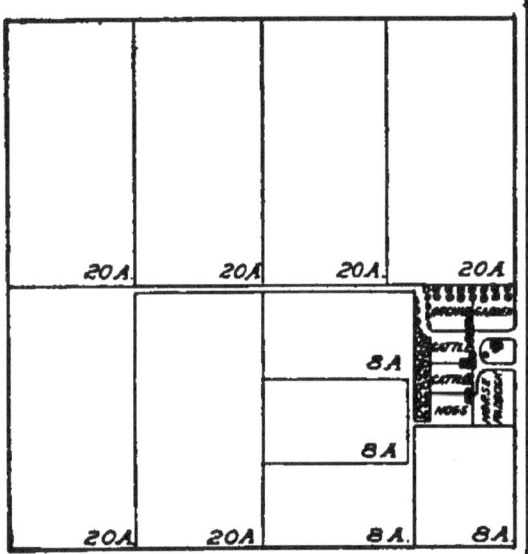

Figure 138.—A 160-acre farm with 8-acre farmstead facing east. Note communication of cattle yards with large fields and hog yards with small fields.

Saving Steps.—The water trough is placed where the stock in the yard can reach it, and also where horses taken out at the front of the barn can also be watered without going out of or into the cattle yard; that is, it extends on both sides of the fence. Beyond the water trough is the machine shed. A team will be taken out of the barn, watered, hitched to a wagon or machine in the shed, and then proceed to the field without going back and forth over the same ground, as would be necessary were the machine shed on one side of the barn and the water trough on the other side. The same saving occurs in coming from the field. The first building reached is the machine shed, where the team will be unhitched. They will then stop at the water trough on the way to the barn.

Yards.—The hog yard is also on the side of the farmstead on which the hog pastures lie. See Figure 138. The cattle yards connect with the main barn, and likewise with the large field. Two yards are provided, so that two different kinds of stock may be fed during winter, and so one may be used during the fall for a stock yard, while the other will care for the cattle. This plan will save the trouble of having to put an extra fence about the stacks every fall.

Waste.—While this plan uses about eight acres of space, every bit of it is used to good advantage and provides an attractive, convenient arrangement for the great amount of work necessarily done on a farmstead.

Questions:
1. Study Figure 137. Draw from memory a sketch of the farmstead and properly locate buildings, yards, etc.
2. Give one or more good reasons for the location of the poultry house, garden, and orchard, as they are in this plan.
3. Why is the machine shed placed where it is instead of on the other side of the barn?

Arithmetic:
1. If one makes three trips each day to the poultry house, how much farther will one walk in a year than necessary. If the poultry house is 5 rods farther from the house than it should be?
2. If a man makes five trips each day to a corncrib, how much extra traveling will he do in a year, if the corn crop is 2 rods farther away than it should be?
3. If a man takes a team to a water trough six times each day, how much does he save in a year, if he saves 2 rods each trip?

CHAPTER XXIII

FARM MANAGEMENT

STANDING OF THE FARMER

Qualification.—To be a successful farmer one must have as much knowledge and ability as a successful merchant, banker, manufacturer, or any other business man. Not long since it was believed that, if one were not intelligent enough to do anything else, one could farm. This belief might have been true, to some extent, many years ago, when the land was newly settled, the soil was rich, and there was practically no market for anything but wheat, so that the chief requirements of a farmer were to plow, to sow and to reap. The more able farmers were more successful than the others, but even the careless and the thoughtless succeeded fairly well as long as the land was rich and free. These crude and careless methods of farming and the number of farmers who used little system or good business management caused farming to be regarded as a rather inferior calling.

Conditions Different.—Conditions are now very different. The fertility of the soil is often somewhat depleted, so that careful and well-planned systems of cropping, tillage and fertilization must be employed to secure good crops. The price of land has increased until one must pay from $5,000.00 to $25,000.00 for a 160-acre farm that at one time could be had free. The country is more thickly populated, and systems of transportation are better, so that nearly any product raised can be marketed. The conditions tend to raise the requirements of the farmer. He must be wide awake, intelligent and ever on the alert for better methods of production. Thus the proper management of a farm demands as high a degree of intelligence as is needed in other walks of life, and we now find as strong, as intelligent and as well educated men and women on the farms as in town. No one is now ashamed to be called a farmer. And an intelligent and successful farmer stands as well in any community as a successful banker, business or professional man. It has been estimated that for the year 1914 the income of American farmers was ten billion dollars. Farming is not only the most important, but the biggest business in existence. The farm property of this country is valued at $32,500,000,000.

Home conditions on the farm may now be fully equal to those anywhere. Science has discovered the septic tank which provides sanitary means for sewage disposal without sewers. Modern machinery has made a water supply under pressure, lighting and heating quite as easily secured as in the city. Rural telephones, rural mail delivery, parcel post and the automobile have brought the country home in as close touch with the outside world as the suburban home. Farm life to-day offers as good opportunities for living, developing and enjoying life as are offered in any city. Most men of wealth own farms and many are building homes in the country to which they hope to retire.

Bright Men and Women Seek the Farm.—The changed conditions are placing more and more bright and progressive men and women on our farms, who study agriculture and put their best efforts into it. As a consequence, we have agricultural schools and colleges, agricultural papers and magazines, and are developing a true science of agriculture. We no longer depend upon "chance" or "good luck" for results in farming, but know the conditions that are necessary to good results and plan and study to supply these conditions. Any boy or girl may well be proud of having been born and raised on a farm and educated for the business of farming.

Questions:
 1. Why did it require less thought and intelligence to farm many years ago than it does now?
 2. Why was farming regarded as an inferior occupation?
 3. What has raised farming to as high a degree as any other occupation?
 4. What is the result of this good standing for the farmer?

Arithmetic:
 1. In 1870 A took a 160-acre homestead. The land cost him nothing. He built $300 worth of buildings. He had 4 horses worth $75 each, 2 cows worth $30, 2 hogs worth $7.50 each, 20 chickens worth 25c. each and $200 worth of machinery. How much had he

invested? How large an income must he have gotten to pay 6% interest on his investment?

2. In 1915 A's farm is worth $100 per acre; he has $800 worth of horses, $1,000 worth of cattle, $150 worth of hogs, $50 worth of poultry, and $1,000 worth of machinery. How much has he invested? What income must he receive to pay 6% interest on his investment?

ROTATION OF CROPS

Definition.—Rotation of crops means changing the crops year after year on a field, so that no field grows the same kind of a crop for several years in succession.

If a field were sown to wheat one year, barley another year and clover another, we would say that the crops are being rotated on that field.

Good farmers are practicing more or less some plan of crop rotation on their farms, because they have found from experience that their fields yield better, if the crops are changed about, than when one field raises the same kind of crop for several years in succession.

Larger yields mean larger incomes. Larger incomes enable farmers to have better homes, better schools, better roads; in short, to live better in every way. We should, therefore, be interested in knowing more about conditions which have a tendency to increase yields.

Systematic Rotation.—A still better practice, because it usually results in larger yields and more profit, is to rotate crops in some regular order, so one may know several years in advance what crop will be grown on each field. Such a plan of cropping is called a systematic rotation. A very simple form of a systematic rotation is to divide the tillable land into three fields of about equal size and crop them as shown in the following chart:

	Field A.	Field B.	Field C.
1st. year	Oats.	Clover.	Corn.
2nd. year	Clover.	Corn.	Oats.
3rd. year	Corn.	Oats.	Clover.

From the above chart one will see that each field has a different crop on it each year for three years, but that the farm is producing the same crops each year. That is, one field is in corn, one in oats and one in clover. In this rotation corn always follows clover, oats always follows corn and clover always follows oats. If a person is practicing such a rotation he can tell, as many years ahead as he wishes, to what crop a certain field will be planted. This enables him to plan accordingly. He knows how many acres of corn, oats and clover he will have each year, and about how many bushels or tons he can ordinarily expect. Thus he may provide the proper amount of storage room for his crops. He will know about how much stock he can keep each year, and can have just what machinery he needs.

Effect on the Soil.—Another advantage gained by following such a rotation is that each crop leaves the soil in good condition for the crop that is to follow. If the corn crop is well cultivated, the oats may be sown the following spring without plowing the land—simply by disking the surface and making a good seed bed. The clover seed is sown with the oats, and makes the crop the year after without extra seeding. Thus three crops are grown and the ground is plowed but once. That is, it is plowed for the corn, grows the corn crop, then a crop of oats and a crop of clover before it is again plowed.

Rotation also helps to keep fields free from weeds. If the corn crop is well cultivated many weed seeds are given a favorable chance to grow, then the next cultivation kills them. The clover crop, as you remember, is cut in the latter part of June, which is earlier than most weeds ripen their seeds. The crop is usually cut again for hay or seed, so weeds are practically given no chance to ripen seed.

All these advantages are gained simply by having a systematic rotation plan to follow, and without increasing the amount of work. But it does require a little more thought than a haphazard system of cropping.

Questions:
1. What is meant by rotation of crops?
2. What is meant by a systematic rotation?
3. Show and explain the cropping of a farm in a simple systematic rotation.
4. What are the advantages of a rotation?

Arithmetic:
1. If but one third of the land is plowed each year when cropped to a 3-year rotation, how much is saved on a farm, with 90 acres of field, if it costs $1.50 per acre to plow?
2. If it costs $9.50 per acre to raise a crop of hay and $12.02 per

acre to raise a crop of oats, how much less does it cost to raise 30 acres of hay than to raise 30 acres of oats?

3. If the rotation of crops adds 15 bus. per acre to the corn crop, how much will it increase the yield on 35 acres? How much is the increased yield of corn worth at 35c. per bushel?

CLASSIFICATION OF FIELD CROPS

Three Classes.—There are a great many different systems of rotation. Some are good and some are not good, and it is well to study a few of the principles, that we may easily know whether a certain rotation is likely to give good results or not.

We rotate crops in order to get good yields, hence we must know something about the effect of each crop upon the soil, so that we may know in what condition the soil will be left for the next crop.

To simplify a study of the general field crops, they may all be placed in three classes, grain crops, grass crops and cultivated crops, basing the classification on the effect each class has on the soil.

Grain Crops.—Under grain crops we can place wheat, oats, barley, flax, speltz, millet and other crops that grow but one year from one seeding and are not cultivated while they are growing. This class of crops has a comparatively small root system, and, as a consequence, very little vegetable matter is left in the soil when the crop is removed. From the time they are sown until they are ripe is usually long enough to allow many of the worst weeds to ripen seeds, so these crops have a tendency to make land more weedy year after year. Grain crops are likely to be sold from the farm, thus removing a large amount of fertility.

Grass Crops.—Under grass crops we can place timothy, bromus, blue grass, redtop and all the common clovers and alfalfa. From the standpoint of the botanist, the clovers and alfalfa are not grass crops, but for our classification we will regard them as such. Grass crops are principally used for hay or pasture. They all grow two or more years, and are not cultivated during their growth. They develop heavy root systems, and, therefore, add much vegetable matter to the soil. Grass crops are usually harvested two or more times during the year, and each crop grows so

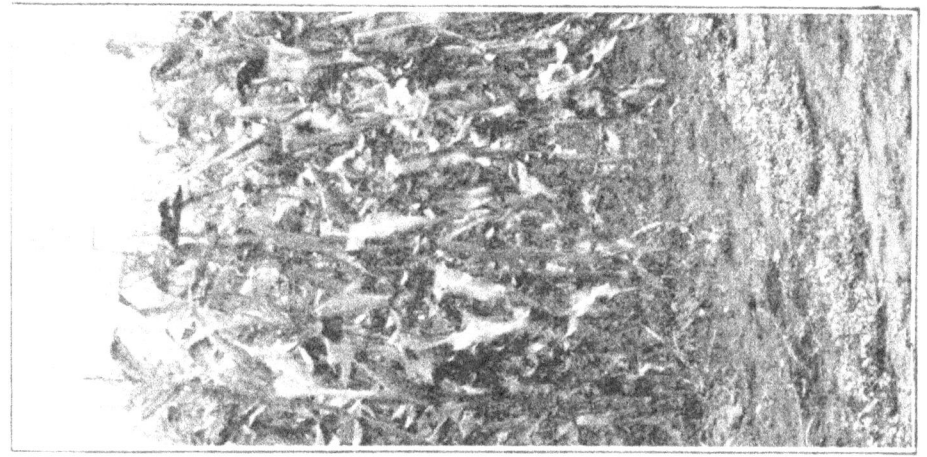

Figure 141.—Corn, classed as a cultivated crop.

Figure 140.—Clover, classed as a grass crop.

Figure 139.—Wheat, classed as a grain crop.

quickly that few weeds have a chance to ripen seed. This crop is usually fed to stock on the farm, and a large part of the fertility removed by the crop is returned in the form of manure.

Cultivated Crops.—Under cultivated crops we can place corn, potatoes and root crops; as mangels, beets, etc., or any crop planted in rows and cultivated while growing. The cultivation destroys many weeds, conserves moisture and causes the liberation of plant food; in fact, on very rich soil much more plant food is liberated than is used by the crop, and much is washed out. On this account cultivated crops are much more exhaustive of soil fertility than the other classes of crops. The corn and root crops are usually largely fed on the farm; hence most of the fertility removed by them is returned in the manure produced.

A Good Rotation will usually include one or more crops from each of these three classes. A rotation of wheat, barley and clover, is not a very good rotation, because it includes no cultivated crop to clear the land of weeds and to conserve moisture. A three-year rotation of corn, oats and clover, has one crop from each class and is a very good rotation for many farms.

A rotation is usually called a three, four, five, six or seven-year rotation, according to the number of crops included in it.

The kind of rotation suited to any farm must be worked out to provide the grain, feed and pasturage needed on the farm. It is possible to plan a good rotation for any farm, but it requires considerable knowledge and ability to select the very best one. To properly manage a farm requires as much knowledge, ability and judgment as to properly conduct any other business.

Questions:
1. What is gained by classifying field crops?
2. What do you understand by grain crops, and what effect do they have on the soil?
3. What do you understand by grass crops, and what effect do they have on the soil?
4. What do you understand by cultivated crops, and what effect do they have on the soil?

Arithmetic:
1. If a bushel of wheat removes 27c. worth of fertility from the

soil, how much is removed from an acre yielding 20 bus. of wheat?

2. If a crop of clover yielding 2 tons per acre removes 100 lbs. of mineral elements worth 5½c. per pound, what is the value of fertility sold in two tons of clover hay?

3. If a bushel of corn removes 14c. worth of fertility from the soil, how much is removed from an acre yielding 50 bus.?

ROTATION MAINTAINS VEGETABLE MATTER

Worn-out Farms.—One often hears the remark that a certain farm is worn out or exhausted. The remark is usually based on the fact that the farm in question no longer yields good crops. Many farms that formerly yielded from twenty to thirty bushels of wheat per acre now yield but from five to ten bushels. When we consider the fact that some of the land in Europe has been cultivated for centuries and is still producing large yields, we must conclude that it is much better, or that farms in this country that are giving such low yields are not exhausted but are simply not in good condition.

Our faith in the latter conclusion is strengthened by the fact that most of us know of instances where men have taken these run-down farms, and after a few years have been able to get as large yields from them as ever.

From the facts at hand, we must conclude that very likely the reason these farms are at present unproductive is because conditions are not favorable for the liberation of plant food. That is, the plant food may be present, but is not soluble so that plants can get it easily. We have learned that the most common way in which plant food is made soluble is by the decomposition of vegetable matter. We may believe, then, that the reason many farms are unproductive is because some of the conditions necessary for decomposition are lacking.

Lack Vegetable Matter.—The condition we are most likely to find lacking in these unproductive soils is a sufficient supply of vegetable matter. A good rotation of crops provides for maintaining the supply of vegetable matter in the soil, because it provides that each field shall grow a grass crop one or more years in every three to six years. Grass crops have heavy root systems and, therefore, add vegetable matter to the soil.

One can get a good comparison of the amount of vegetable matter added by grain crops, and by grass crops, by pulling a few of the stubbles of grain and of clover and timothy. One will find that the grain stubble is pulled up very easily and that some digging must be done to get stubble of any of the grass plants.

In the simple three-year rotation, corn, grain and clover, discussed on page 308 you will see that each field will be in clover once every three years. This will add sufficient vegetable matter to keep the soil reasonably well supplied during the other two years while the corn and grain crops, or crops that exhaust the vegetable matter, are growing.

Cultivated crops, as corn, probably return as much vegetable matter to the soil in the roots and stubble as do the grain crops, but they use during their growth a great deal more than the grain crops, on account of the cultivation given them. We have learned that moisture and air are essential to decomposition. By the cultivation of these crops the soil is loosened on top, so that air can enter freely and so that moisture does not rise to the surface to evaporate, as it rises in a grain field that is not cultivated while the crop is growing. Thus the two essentials to decomposition, air and moisture, are maintained. Consequently decomposition is more rapid and more of the vegetable matter is decomposed.

A rotation which provides for growing grass on one third of the tillable land furnishes a great deal of pasture and hay, which usually is an incentive to keep more live stock. Hence more manure is produced than is usually available on a grain farm. More manure can be applied to the fields; and this, too, has a marked effect in keeping up the supply of vegetable matter.

Questions:
1. What is generally lacking when a soil ceases to be productive?
2. In what ways does the rotation of crops tend to increase the supply of vegetable matter in the soil?
3. How can you prove that grass crops add more vegetable matter to the soil than grain crops?

Arithmetic:
1. It costs $13.04 per acre to grow a crop of oats. How much will it cost to grow $33\frac{1}{3}$ acres of oats?
2. It costs $16.20 per acre to grow a crop of corn. How much will it cost to grow $33\frac{1}{3}$ acres of corn?

3. It costs $9.50 per acre to grow a crop of mixed hay. How much will it cost to grow 33⅓ acres of mixed hay?

4. It costs $13.07 per acre to grow a crop of wheat. How much will it cost to grow 100 acress of wheat?

5. Using the figures given in the preceding examples, how much more does it cost to grow 100 acres of wheat than to crop 100 acres in a rotation with ⅓ in corn, ⅓ in oats and ⅓ in hay?

PLANNING FARMS

Farm and Farmer.—In the last few pages we have learned something about the rotation of crops, its effects on the soil, and how to tell whether or not a certain rotation would be likely to keep the soil in good condition and give good yields. Before we can plan a suitable rotation for any particular farm we must know certain facts about the farm and the farmer.

Sketch of Farm.—We should have a rough sketch of the farm in question, showing its shape and size, the location of the farmstead (farmstead includes buildings, yards, orchards, garden, drives and lawn), the size and shape of the fields and pasture, the fences and lanes and the sloughs and waste places. We should know also the kind of soil, the location of the farm, the amount and yields of the different crops grown, the markets, and the ability and desires of the farmer.

The Farmstead.—The location of the farmstead determines the distance each field will be from the base of operation, the distance live stock will have to go to pasture on the different fields, the amount of lane necessary to reach them, and whether or not they must be driven across a public road or a railway track.

The size and shape of the fields will determine the size and kinds of machines that may be used and the type of farming to be done. If there are only a few small and irregular fields, one cannot grow grain to advantage and compete with farmers who have large, straight, level fields. On the few small fields the farmer would need to grow some crop he could care for to advantage with small machines, and he would also want to grow some crop that would bring in considerable per acre. Five to ten acres of grain or corn would not produce an income large enough to support a family; but five or ten acres in small fruit or vegetables would insure products of sufficient value to make a good income and provide labor for a fair-sized family.

Fencing.—The amount of fencing on a farm and the cost per acre of fencing the various fields are factors which would influence one in deciding on the kind and amount of live stock to be kept.

Waste land, or land which for some reason cannot be cultivated with the regular fields, must be considered in

Figure 142.—It costs much more per acre to plow a small irregular field like the above, than a long field as shown in Figure 143.

Figure 143.—Plowing a long, regular field. Compare with Figure 142. Which field would you prefer to plow?

planning a rotation. If there is any considerable amount of such land that can be used only for hay or pasture, stock must be provided to utilize this feed, and very likely a smaller proportion of the tillable land would be needed for hay and pasture, so that a correspondingly larger acreage could be devoted to other crops.

Soil and location determine the kind of products that can be raised and the kinds desirable to raise. One must not plan a rotation providing for a large acreage of corn, in a community or on a soil not adapted to corn production. Likewise it would not be well to plan to raise a heavy bulky crop, like potatoes, where one is a long distance from market, or to keep dairy stock where facilities are poor for marketing dairy products, or to keep beef stock or hogs where grain feed, as corn or barley, is difficult to grow.

The acreage and yields of the different crops that have been grown are good indications of the type of farming carried on in the community, the condition of the farm and the kind of farming with which the owner is familiar and best adapted to do.

The ability and desires of the farmer are probably the most important consideration. If a man does not like live stock or a certain kind of farming, it is very likely that he will not succeed with it, though the plan of managing the farm might be excellent. Likewise a farmer might have ability as a market gardener or in raising horses or sheep, but might fail at dairying or general farming.

Questions:
1. What are some of the facts we should know about a farm to enable us to plan a suitable rotation?
2. What does the location of the farm determine?
3. What should the size and shape of the fields determine? The amount of fencing? The waste land? The soil and location? The size of the farm and yields?

Arithmetic:
1. If one raises 10 acres of wheat yielding 20 bus. per acre, how much is the crop worth at 85c. per bushel?
2. If one raises 10 acres of potatoes yielding 150 bus. per acre, how much is the crop worth at 60c. per bushel?
3. If one raises 10 acres of onions yielding 400 bus. per acre, how much is the crop worth at 60c. per bushel?

ARRANGEMENT OF FIELDS

A Map.—If one were to make a map of the average farm, showing all the fields as the farm is cropped for one year, one would be likely to be surprised at the number of fields and their irregular shape. Often a little careful planning will result in a great saving of time and labor.

Fields for a Rotation.—If the rotation of crops is to be practiced, the farm must be divided into as many fields as there are years in the rotation or rotations. To do this in a way to be most economical of fences and to get the fields properly located, necessitates a careful study of conditions. In arranging the fields of any farm it is desirable to

(1) Have fields of uniform size.
(2) Have fields of convenient shape to work.
(3) Have one end or each field as near to the farmstead as possible.

318 ELEMENTS OF FARM PRACTICE

(4) Economize in fencing.

(5) Make the best use of all parts of the farm.

Uniform Size.—Fields should be of uniform size to make the farming business systematic, so a like amount of various crops may be grown each year. This method regulates the amount of labor and machinery needed and of live stock that may be kept, and makes possible a really systematic arrangement of the farm business.

Shape.—As a rule long fields are more desirable than square ones, as machines can be used on such fields to better advantage, but this suggestion must not be overdone, especially if the fields are to be fenced, for long fields require more fencing per acre than shorter ones.

Distance from Farmstead.—A great many trips must be made each year to each field and a difference of several rods in distance from the farm buildings makes a great difference in a year or in a lifetime. Figure 144 illustrates a 160-acre farm in Minnesota, (a) as the fields were arranged, and (b) as they are planned for a systematic rotation. This whole farm is tillable. A study of these two plans shows that by the rearrangement of fields there will be a saving of 252 rods of fencing, if all fields were to be fenced. In the new plan, the fields are of excellent shape to work, are all the same size, and the average distance of the fields

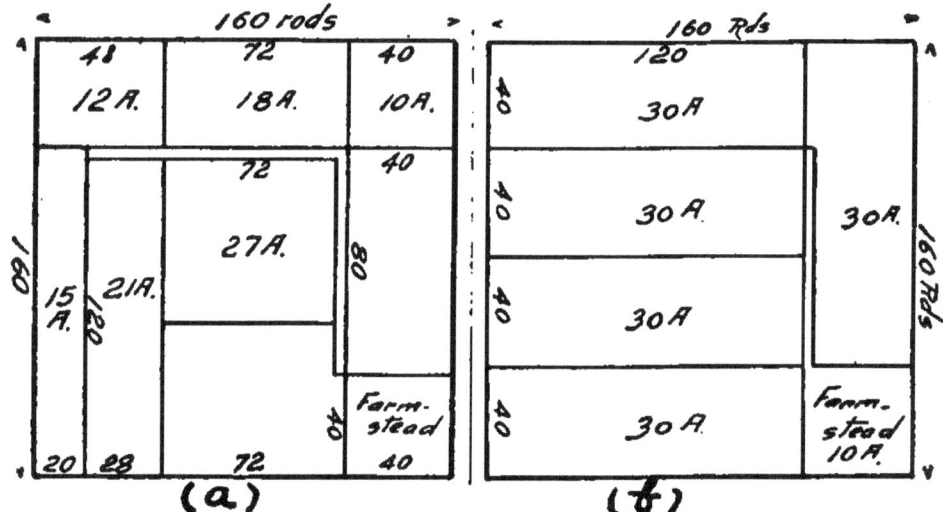

Figure 144.—A 160-acre farm. (a) Fields poorly arranged. (b) Fields well arranged. Note amount of fencing required to enclose all fields in (a) and compare with (b).

from the farmstead is 24 rods, while in the old plan the average distance of the fields is 70 rods.

Economize in Fencing.—The size and shape of fields have much to do with the amount of fencing required to enclose an acre. Figure 145. See arithmetic lesson below.

If one wishes to divide an 80-acre farm into five fields of equal size, there is room for study. The proper solution

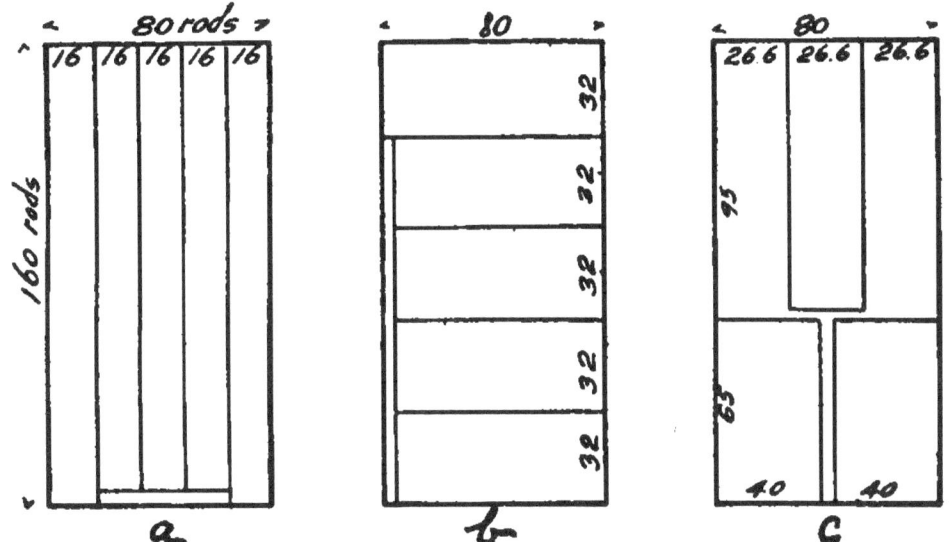

Figure 145.—An 80-acre farm divided into five fields in three different ways. Figure the amount of fencing required to enclose the fields in each case.

may mean quite a saving in fencing and in operating the fields. The three plans, a b, and c, in Figure 145, illustrate three ways of dividing an 80-acre farm into five equal sized fields. An 80-acre farm is usually 80 rods wide and 160 rods long. If it is divided as shown in (a), 640 rods of fencing would be required for the inside fences. If divided as shown in (b), 448 rods would be required, and if divided as shown in (c) only 426 rods would be required.

Questions:
1. What are the advantages of having fields uniform in size, if one practices a rotation of crops?
2. Can a farm be so divided as to make any difference in the average distance of fields from the farmstead?
3. What effect does the shape of fields have on the amount of fencing required per acre to enclose them?

Arithmetic:
1. How many acres of land in a field 1 rod wide and 160 rods

ELEMENTS OF FARM PRACTICE

long? How many rods of fencing are required to enclose it? (160 square rods in an acre.)

2. How many acres in a piece of land 16 rods by 10 rods? How many rods of fencing are required to enclose it?

3. How many acres of land in a field 40 rods square? How many rods of fencing are required to enclose it? How many rods of fencing are required per acre?

4. How many acres in a field 80 rods square? How many rods of fencing are required to enclose it? How many rods of fencing per acre?

A PRACTICAL ROTATION

Application.—To apply the principles of crop rotation and farm planning to an actual farm, we will study one on which a good system of rotation has been practiced for several years.

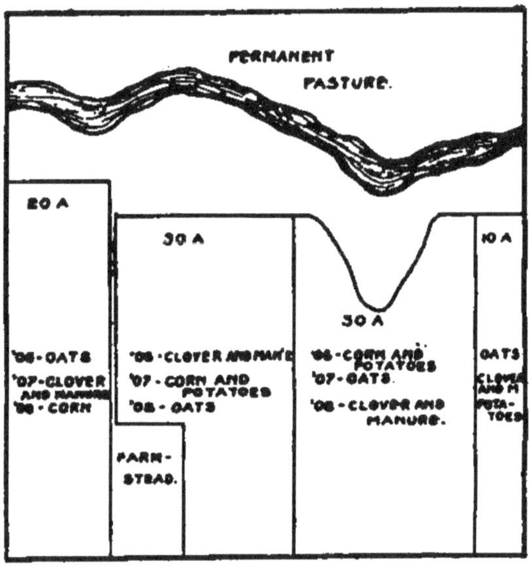

The farm we will take is a 160-acre farm, one and a half miles from a good town. The soil is light, sandy loam. The owner specializes in growing potatoes and live stock. He is a very careful farmer, a study of whose methods and farm will be valuable.

The accompanying chart, Figure 146, shows the lay of the farm and the arrangement of the fields. Nearly half of the farm, the back part, is broken up by a river and bluffs so that it is not tillable and can be used only for pasture. The remainder, or 90 acres, is all well drained, reasonably level and easily worked.

Figure 146.—A 160-acre farm on which a three-year rotation has been successfully followed for a number of years.

The Rotation.—The 90 acres of tillable land is divided into three 30-acre fields, or rather two 30-acre fields and a 10 and a 20-acre field. These two smaller fields are farmed as though they were only one field; so they make the third 30-acre field. One field raises grain, one field raises clover and one field raises corn and potatoes each year, but no field grows the same kind of a crop two years in succession.

The Potato Crop.—The chief field crop that is raised for sale is potatoes; so the whole farm is planned to give good crops of potatoes. The first crop of clover is cut for hay. Sometimes the second crop is cut for seed, but more often it is plowed under to add plant food for the potato crop which is to follow. All the manure made on the farm is also applied to the clover land. The clover and manure have made the land rich and the clover crop has aided in cleaning it of weeds. The potatoes have a clean, rich soil in which to grow.

Corn.—Only a part of the clover sod plowed up is planted to potatoes. The remainder is planted to corn. The corn and potatoes are regarded as one crop in the rotation, since they are both cultivated crops and have about the same effect on the soil. He gives his corn land the same careful preparation that he gives his potato land, and consequently gets good crops.

Oats.—After he has grown his crop of corn and potatoes he wishes to seed his land to clover again, because he has found that it is the best crop he can raise to put his land in good condition for corn and potatoes again; besides, he needs the clover hay for his cows. As clover must be sown with a grain crop, he seeds this corn and potato land the following spring to oats. With oats he sows clover and timothy seed for a crop the year after the oats are harvested.

Clover.—Getting a catch of clover is the key to his success as a farmer; so he takes every precaution to be sure of a stand. The cultivation given the corn and potatos saves moisture, and the manure and clover in the soil give up plant food, so that there is more moisture and plant food in such land the following year than there would be if the field had been pastured or had grown a crop of grain. This extra moisture and plant food help to start the clover seed sown with the oats the year after the land grows corn and potatoes.

Results.—That this kind of farming pays is shown by results. Besides good crops of clover, corn and oats, the potato crop on this farm brings from $60.00 to $100.00 per acre annually. As it costs about $30.00 per acre to raise potatoes, it is plain that there is a profit on these crops.

Questions:
1. What is the rotation practiced on the farm discussed above?
2. How is the land prepared for potatoes?
3. What crop follows potatoes and corn in this rotation?
4. What crop follows the oat crop, and when is it seeded?

Arithmetic:
1. If a farmer raises 15 acres of potatoes each year, how many bushels will he have, if the yield is 165 bus. per acre?
2. If Mr. Brown raises 15 acres of corn each year, how many bushels will he have, if the yield is 50 bus. per acre?
3. If oats yield 48 bus. per acre, how many bushels will 30 acres produce?

A FIVE-YEAR ROTATION

Rearrangement of a Farm.—A 160-acre farm in southeastern Minnesota, four miles from a good milk market was cropped in 1904 as shown in Figure 147. This farmer was carrying on general diversified farming, and without changing the type of farming in the least, his farm was replanned and a systematic rotation of crops arranged that would certainly make the farm more attractive, more easily worked, and more productive, if put in practice. In Figure 147 note the small and irregular fields, the distance some of them are from the farmstead, and the lack of system in cropping.

Without materially changing the amount of land devoted to each kind of crop this farm may be rearranged in five uniform fields of convenient shape and size (27 acres each) and one end of each field be as near to the farmstead as is easily possible on such a farm. See Figure 148. The 12-acre field in the southwest corner is too wet to cultivate; so it is left as permanent meadow.

Rotation.—A five-year rotation would be well adapted to such a farm, as it would provide about the same amount of hay and pasturage as was formerly used. This rotation would be corn, grain, meadow and pasture. That is, one field would produce corn, two fields would produce grain, one would produce hay and one would produce pasture each year. The field that grows corn the first year would produce grain the second and third years, meadow the fourth year, and pasture the fifth year.

The Grain Crops.—The first grain crop after the corn would be sown on the corn land, usually without plowing but simply disking it well so as to make a good seed bed. After this grain crop was harvested the land would be plowed in the fall, so that it would have time to settle down and become compact by spring. The next spring it would be sown to grain again, but, with the grain, grass seed, timothy and clover would be sown to make the crop for the two years following.

Meadow.—We have learned that grass crops as meadow and pasture are beneficial to the soil, as they clean it of

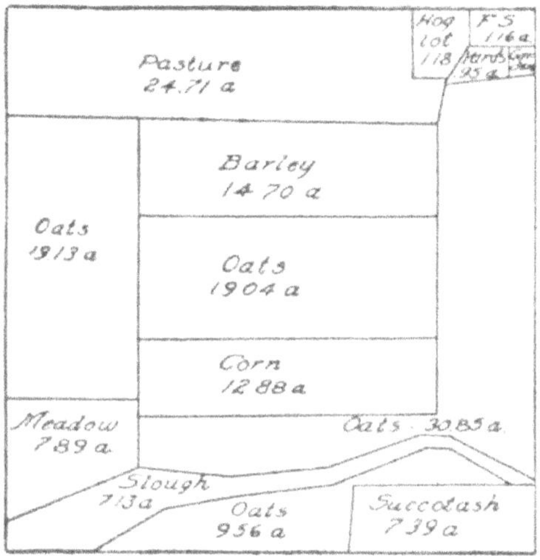

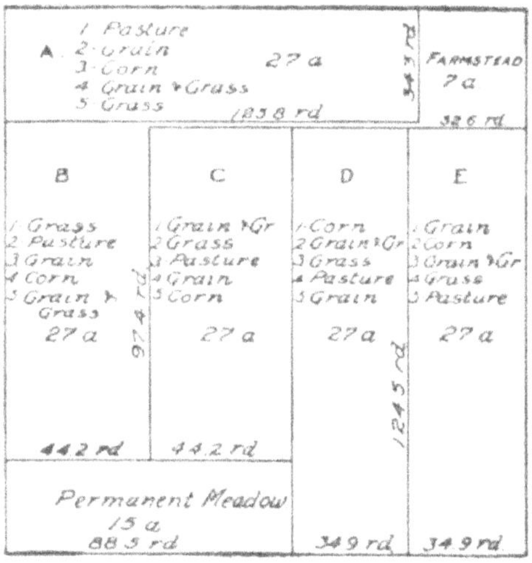

Figure 147.—A 160-acre farm in southeastern Minnesota, cropped in 1904. Compare with the reorganization plan in Figure 148.

Figure 148.—The 160-acre farm reorganized for a five-year rotation. Note convenient shape and arrangement of fields, and that there is little change in the acreage of crops grown. The fields are simply arranged better and a systematic rotation planned.

weeds and add vegetable matter. This rotation provides for having each field in grass two years out of five. The first year the grass would be cut for hay and the second year it would be pastured.

Pasture.—Pasturing land occasionally, as provided in this rotation, is beneficial to the soil, as practically all the crop grown during the year is left on the field as manure, and the development of the roots adds vegetable matter also. Pasturing usually puts land in good condition for other crops. One can haul manure upon it during the summer, when the other fields are growing crops. This manure plowed under with the pasture sod makes a good seed bed for corn.

Corn.—In this rotation corn is planted on manured pasture sod each year. This is a very desirable place for corn, and usually results in good yields.

Study the accompanying charts and, if possible, draw them and put in the crop that would be grown on each field each year. The figures 1, 2, 3, 4 and 5 indicate the 1st, 2nd, 3rd, 4th and 5th years of the rotation. Each crop is produced each year.

This five-year rotation is a good one for many farms, and every farm boy ten years old and over should thoroughly understand it and its advantages over no rotation.

Such a rotation tends to keep the fields clean of weeds, productive, and to economize labor. Each field is plowed but twice in five years and seeded but three times in the five years. Still it is kept in good condition for the crop it is to grow, because the crops are so arranged that each crop helps to fit the soil for the one that follows.

Questions:
1. What is the five-year rotation described above?
2. If one had five 20-acre fields cropped to the above 5-year rotation, how many acres of grain would one raise each year? How many acres of corn? Of pasture?
3. In what ways does the five-year rotation described above aid in cleaning the land of weeds?

Arithmetic:
1. If a farmer on the above 160-acre farm kept 15 cows, 8 two-year-olds, 10 yearlings and 12 calves, how many head of cattle would he have?
2. If each cow required 1 acre of pasture for the summer, each two-year-old $\frac{3}{4}$ acre, and each yearling $\frac{1}{2}$ acre, how many acres of pasture would be needed?
3. If each cow required, during the winter, $2\frac{1}{2}$ tons of hay, each two-year-old 2 tons, each yearling 1 ton, and each calf $\frac{1}{4}$ of a ton, how many tons of hay would be needed?
4. If the 27 acres of rotation meadow produced two tons per acre, and the 12 acres of permanent meadow produced $1\frac{1}{2}$ tons per acre, how many tons of hay would the farm produce?

FARM ACCOUNTS

Bookkeeping.—One of the first essentials of successful farm management is a set of accounts that will show which enterprises are paying and which are not. To keep a complete set of books for all the enterprises on a farm requires considerable time and training, but most any farmer or boy or girl can, with little effort, keep a reasonably accurate account of one or more of the leading enterprises on his own farm. We will not try to give a complete system of farm bookkeeping, as too much space and study would be required, but we do wish to study with our readers some of the problems of farm management affecting a few of the more general enterprises of the average farm, and to show the application of accounts and figures to farming as a business, and the value of their use.

Enterprises.—Some of the main enterprises on the general farm are: Horses, Cattle, Sheep, Hogs, Poultry, Fruit, Grain, Hay, or, in short, any of the various classes of products produced. If a farmer keeps only a cash account showing the receipts from products sold and the amount spent, while he might make a profit on the whole farm, he might lose on sheep, hogs or some of the other enterprises, and not know it. If he had an account with each enterprise, he would know on which he was losing and on which he was making a profit, and could plan his future work so as to increase the profitable lines and decrease or improve the unprofitable ones.

The Farmer as a Merchant.—The farmer buys and sells products just as truly as the merchant. The way be buys most of the products he sells is somewhat complicated. He seldom knows, without accounts, just how much any particular thing, as a hog or bushel of grain, has cost him. The cost of a product of the field to a farmer includes rent on land, seed, man and horse labor in preparing the land, seeding, cultivating, harvesting, machinery cost; also any cash expenditure, as for twine, oil or threshing. The cost of a live stock product includes labor, feed, shelter, interest on investment and depreciation. It is not difficult to keep a fairly accurate account of some leading enterprises, and it gives a much better grasp of business than can be gotten in any other way.

Land Rent.—Land rent is one of the items of cost in crop production that must be considered. This cost is very evident if one does not own the land, but pays rent for the use of it. If the farmer owns land himself he must charge a fair rent per acre against each crop, because he expects his money invested to be worth a certain rate of interest. That is, he could loan his money at a fair rate of interest, if it were not invested in the farm. If he did not consider rent as an item of cost against each crop, he might apparently make a profit in his farming operations while in reality he was losing—that is, growing crops at a loss, but making more than wages on his labor on account of the income on his investment. In such case, he had better sell his farm and loan his money at 5% or 6% interest.

Questions:
1. For what reasons should a farmer keep accounts?
2. In what way does a farmer buy the hogs, milk, etc., that he sells from the farm?
3. Why is land rent one of the items in the cost of production, even though a man may own his land?

Arithmetic:
1. If a farmer feeds a cow each day for 200 days 25 lbs. of hay worth $5.00 per ton and 6 lbs. of grain worth $20.00 per ton, what is the cost for feed for the 200 days?
2. If pasture for a cow is worth $1.00 per month, what is the cost of pasturing a cow 165 days?
3. If ½ hour per day is required to care for a cow during the 200 days she is kept in the barn, and ¼ hour per day for the 165 days she is pastured, how many hours of labor are required to care for a cow a year? What is this labor worth at 18c. per hour?
4. What is the total cost of feed and labor for the cow for the year? How many pounds of butter-fat must she give in the year to pay for these items, if butter-fat is worth 30c. per pound?

LIVE STOCK ACCOUNTS

Making Work Interesting.—Nearly every farm boy, while he is attending school, has more or less work with the stock at home. This work done mornings and evenings, often under unfavorable conditions, is sometimes uninteresting to say the least, though it is as valuable a part of one's education as are the things one learns at school. Keeping records with the various classes of stock, so that one knows which class is giving the best returns for feed and the highest price per hour for labor, adds to the work an interest that can be gained in no other way.

In order to know the profits or losses of an enterprise, certain charges must be made and deducted from the value of the product. The main charges against live stock are: depreciation, interest on investment, feed, labor, shelter.

Depreciation is the difference between the value of any property at the beginning and at the end of the period during which an account is kept of it. For example, if, at the beginning of a year, one has ten cows worth $400, and at the close of the year they are worth less or he has lost or sold some, so that the total value of cows on hand is but $350, there will have been a depreciation in value of $50,

Figure 149.—A fine bunch of hogs. The questions one should be able to answer after producing a lot of hogs are: How much did they cost per pound? How many pounds of corn or other feed did it take per pound of pork, etc.

which must be accounted for in the charges against the stock. Likewise there may be a gain in the value of the stock for a given period. If so, it must be credited to the enterprise. This loss or gain is most easily accounted for by taking an inventory at the beginning of the year—that is, making an estimate of the value of the stock on hand and charging the enterprise with this amount, then crediting the enterprise with the inventory value at the close of the year.

Interest on the investments must also be charged against an enterprise, if an accurate knowledge of loss or gain is to be had; because, if the money invested in stock were loaned, it would earn a certain rate of interest. One would not care to invest money in stock if one could not get as much interest on it as if it were invested in some other enterprise. A very common rate of interest is 6%. Thus, if one had $400 invested in live stock for a year, one of the charges against the stock would be an interest charge of $24, as that is the amount $400 would earn, if loaned at 6%.

Figure 150.—Weighing hay. If one weighs hay a few times when feeding one will soon be able to tell approximately how much one is feeding without weighing it every time.

Feed.—The value of all feed consumed must be charged to the stock. This is usually the only charge considered, but it is evident that the other items mentioned are as legitimate charges. It seems at first thought that it would be difficult to keep account of the amount of feed fed to each cow or to all the cows, but very little time is required to get approximately the amount fed for a month. If one carefully weighs, for a few days, the hay and grain that are fed, one can soon learn to feed about the desired amount without weighing, but simply by using the same measure for grain and giving about the same sized forkful of hay or the same number of bundles of fodder. When one knows about the amount of feed fed per day, the amount fed per month can be ascertained by multiplying it by the number of days in the month.

Labor or any work done in caring for any class of live stock or marketing the product must be charged against the enterprise; because, if a man hires labor, he must pay for it, or, if he does the work himself, he is entitled to wages. With a little thought, one can determine about the amount of time required each day to care for any class of live stock; and, by multiplying the amount by the number of days in the month, can get the amount of work done in the month.

Cost of shelter is not so easy to determine; yet it is an actual cost against the live stock. An easy way to get the approximate cost for shelter, is to find out the value of the building or part of the building in which the stock is kept, then figure 8% of this value as the annual cost of shelter. The 8% will allow for interest, insurance, taxes, repairs and depreciation.

Questions:
1. What will add an interest to caring for stock?
2. What are the main charges to be made against live stock?
3. Explain each charge.

Arithmetic:
1. On Jan. 1st, 1914, a barn is worth $1,000. On Jan. 1st, 1915, it is worth $950. How much has it depreciated in value? How long will it last, if it depreciates the same amount each year?
2. If a $50 cow lives ten years, what is her annual depreciation?
3. If a barn that shelters 40 head of stock costs $2,000, how much is the annual cost of shelter, if one figures 8% on value of barn? How much is the annual cost of shelter per animal?

AN ACCOUNT WITH A COW

Actual Figures.—To simplify the account with the dairy stock and to illustrate what any boy may do at home with some cow he is milking and caring for, we will use a record which shows the average results per cow for a year in a herd of 14 cows of which an accurate record was kept.

A Business Statement Showing the Cost and Income of One Cow for the Year 1908.

	Dr.	Cr.
Int. on investment at 6%	$ 2.40	
Value of grain fed	10.46	
Value of roughage fed	12.29	
Value of pasturage	5.00	
Cost of labor	23.28	
Cost of shelter	3.20	
Miscellaneous expense	1.50	
Net profit	.43	
Income for year		$58.56
	$58.56	$58.56

(Note.—The 43c. net profit was obtained by deducting the sum of the seven items of expense, $58.13, from the $58.56 income.)

The foregoing is a complete business statement except the opening and closing inventory. These were left out to simplify the account. To put them in, one would simply put in the debit column the value of the cow at the beginning of the year and in the credit column her value at the close of the year.

Interest on Investment.—To get this item we simply assumed the cow to be worth $40.00 and figured 6% on this amount.

Feed.—The cow was fed in the stall for seven months, during which time she ate 301 lbs. of farm grain and 442 lbs. of mill feed, worth $10.46, and 1,496 lbs. of hay and 3,330 lbs. of fodder, worth $12.29. She was pastured for five months and was charged for this at the rate of $1.00 per month, which is the ordinary charge for pasturing.

Figure 151.—Weighing feed.

Labor.—The labor includes all time spent in milking and caring for the cow and in marketing the product. As the milk was shipped, it had to be delivered to the station every day, which work required considerable time for both man and team, all of which must be accounted for.

Miscellaneous Expenses.—The item for miscellaneous expense is the actual wear and tear on dairy equipment, cost of medicine, etc.

Profit.—The net profit appears very low; but in reality it is not bad at all, since every bit of work done and feed fed was paid for in full and a fair rate of interest has been paid on money invested in the cow and in the buildings.

There is an additional profit to the farm by keeping live stock, as most of the fertilizing value of feeds fed is retained on the farm in the form of manure. The manure produced by a cow in one year is worth several dollars to the farm.

The cow also had a calf which is worth something; and, had the butter-fat been sold to the creamery instead of shipping the whole milk, about 5,000 lbs. of skimmed milk would have been available for feed; which is worth, at 15c. per hundred pounds, $7.50.

We hope some of our readers will begin at once to weigh the milk produced and feed consumed by some or all of their cows. Also, keep a record of the amount of time spent in caring for them. It is very interesting to foot up such accounts each month and to know whether one is making or losing by keeping stock.

Questions:
1. What do you understand by the terms Inventory, Depreciation, and Profit?
2. How can you determine the number of hours of labor required to care for a cow a year?
3. What advantages are there in keeping cows besides the profit shown in an account similar to the one given above?

Arithmetic:
1. If a cow gives 18 lbs. of milk per day, how much will she give in 300 days? How much butter-fat does she give each day, if her milk tests 4% fat? How much butter-fat will she give in 300 days? How much is the butter-fat worth at 30c. per pound?
2. If a cow is fed each day 4 lbs. of corn worth 56c. per bushel (56) lbs.), 2 lbs. of bran worth $25 per ton, 12 lbs. clover hay worth $5 per ton, and 10 lbs. of fodder corn worth $4 per ton, how much does it cost to keep her one day? To keep her 200 days?
3. How much does it cost to pasture a cow 165 days at $1 per month?

MARKETING DAIRY PRODUCTS

Item of Expense.—Getting dairy products to market is an item often overlooked in considering the cost of production and the profits in dairying. This item is much larger than one would believe at first thought. It is, however, a necessary item of expense, but often a little consideration and planning will greatly reduce the cost and add a corresponding amount to the profits.

Making Butter on the Farm.—A few cling to the old method of making butter on the farm, and there are probably places where it may be necessary to do so; but, where it can be avoided and the cream or milk sold at a reasonable price, it is preferable to sell it. In churning by hand, more of the butter-fat is lost than when cream is churned in a large churn at the creamery. Butter makers in creameries do nothing but make butter. They make a study of it; and, having better facilities than are usually found in the home, make a better quality of butter.

Overrun.—Milk usually contains from 3% to 5% butter-fat, and cream from 20% to 40% butter-fat. A pound of butter-fat will make more than a pound of butter, because butter contains from 12 to 15% water; also some salt and casein. This increase in weight is called by butter makers the overrun. A good butter maker with modern creamery equipment can get an overrun of from 18% to 24%. If he buys 100 lbs. of butter-fat he can make from 118 to 124 lbs. of butter from it. It is very seldom that one can get as large an overrun, when churning a small amount of butter on the farm, as can a good butter maker in a modern creamery. While a farmer can get more pounds of butter by churning his cream himself than he had pounds of butter-fat in the cream, yet he cannot as a rule get as many pounds of butter as could a butter maker by churning the same cream in a modern creamery. This fact, together with the fact that home dairy butter is not quite so marketable as creamery butter, makes it the part of wisdom, on most farms, to sell the cream or milk at the local creamery; or, if no creamery is convenient, to ship either the milk or the cream, rather than to make butter on the farm.

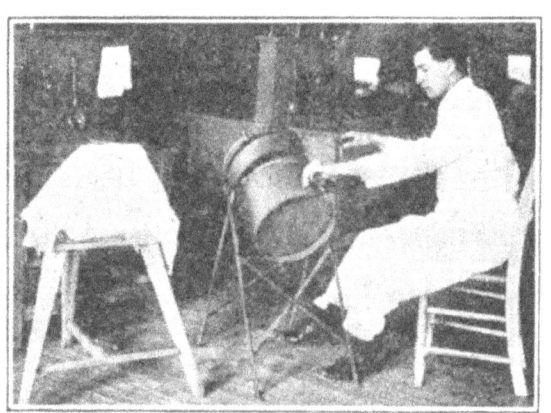

Figure 152.—Home manufacture of butter.

Exceptions.—There are times when it may be wise to make butter on the farm, but at present such conditions are exceptional. In deciding which method to follow one should not overlook the fact that butter making on the farm usually falls to the housekeeper, who, as a rule, has too much to do without this unnecessary work.

Selling Whole Milk or Cream.—The majority of farmers now separate their milk, keep the skimmed milk at home for their calves and pigs, and sell their cream. A few, however, who live within reasonable shipping distances from large cities, ship the whole milk to retail dealers who retail to the consumers. When milk is shipped in this way it is usually sold by the hundred pounds, and it is often a problem to determine which is more profitable—to ship whole milk or to sell the cream. A specific problem will illustrate how any one must decide which is the better method. A farmer living three miles from town, and the same distance from a creamery, has ten cows, each giving daily 20 pounds of milk testing 4 per cent butter-fat.

Whole Milk.—If this farmer ships whole milk to the city, he must deliver it every day, seven days each week. It will take a man and team two hours each day to take the milk to the station. This time is worth about 30c. per hours. Thus the cost of marketing will be 60c. per day or $4.20 per week.

Cream.—If he separates his milk and sells the cream, he will need to deliver it but three times a week, which labor at 60c. a trip will cost $1.80. He will have to separate 100 lbs. of milk and wash the separator fourteen times during the week. Allowing half an hour for this, it amounts, in a week, to seven hours. Man labor is worth about 15c. per hour, which will make the separating cost $1.05. The cost for interest, depreciation and repair on the separator will be about 25c. per week. Thus the separating and delivering will cost $3.10 per week.

In addition to a saving in the cost of marketing, if the cream is sold in place of the whole milk, this farmer will have about 1,175 pounds of skimmed milk, which at 15 cents per hundred is worth $1.76.

Questions:
1. What item of expense is often overlooked in the cost of dairy products?
2. Give at least three reasons why it is usually not wise to make butter at home.
3. What do you understand by the creamery man's term "the overrun"?
4. Compare the items of expense in selling whole milk with selling cream.

Arithmetic:

1. If a butter maker gets an overrun of 20%, how much butter will he make from 986 lbs. of cream?
2. If a man has 10 cows, each giving 20 lbs. of milk per day, how many pounds of milk will he get per day? How many pounds of butter-fat will he get, if there are 4 lbs. of fat in each 100 lbs. of milk?
3. What is the value of 1,175 lbs. of skim milk at 15c. per hundred pounds?

CO-OPERATION IN DELIVERING MILK OR CREAM

Co-operation Reduces Cost.—We have found that the farmer who hauled his cream to the creamery three times each week spent 6 hours of time for himself and team; which, at 30c. per hour, costs him $1.80 per week or $93.60 per year, simply for marketing the cream from ten cows, each giving daily 20 lbs. of milk testing 4% fat. This is certainly quite an item, especially at a time when farm help is as scarce as it is at present.

If three such farmers would co-operate in hauling cream, so that each would haul but one day each week, the cost would be very little more than one third as much as when each markets his own cream. There would be a saving each year of about $60 per farm or $6 per cow.

We have found (page 333) that it cost a farmer, with ten cows, $4.20 per week or $218.40 per year to market whole milk, because he was obliged to go to market every day. If three such farmers, living near each other, were to co-operate in marketing their milk, about ⅔ of this cost, or $155, would be saved to each farm; which is over $15 per cow per year. An increased profit of $15 per cow is worth consideration.

Creamery Company Hauls Cream.—It has been suggested that instead of each farmer's hauling milk or cream to a creamery, the creamery company employ teams to haul the cream or milk from all the farms. It would certainly seem that there might be a very great saving realized, if this suggestion were followed. One man and team thus employed could haul all the cream from twenty to sixty farms, depending on the condition of roads, distance to haul and size of herds.

Consider.—Spend a few moments figuring on these problems as they may apply in your particular locality. If you can answer the following questions, you can figure the problems easily.

How often per week do you deliver milk or cream?

How much time, on the average, is required?

How many hours will a man spend in a year to deliver your milk or cream? What is the total cost of this labor at 15c. per hour?

How many hours of horse labor will be required in a year to deliver your milk or cream? What is the total cost of this labor at 8c. per hour per horse?

These figures will enable you to find the total cost of marketing the product of your dairy when you do it yourself. Figure, also, what it would cost you, if you were to exchange with two or three of your neighbors, so that you will have to go but every third or fourth time.

Still another valuable problem will be to find out how many farms could be reached by one team circling around so as to reach the greatest possible number of farms and get back to the creamery by traveling from 10 to 14 miles.

To make such a trip one half day's time for man and team would be required, at a cost of $1.50 to $2.00. Figure whether or not this would be a saving over the common practice of each farmer's delivering his own cream.

Questions:
1. In what manner are the dairy products of your farm marketed?
2. Are there two or more of your neighbors living near your place, so that you might co-operate with them in hauling your milk or cream to the creamery or to the station?
3. Would it not be practical for the creamery company to employ one or more teams to collect milk or cream, instead of having each farmer deliver it?

Arithmetic:
1. If 100 lbs. of milk were run through a separator and 4 lbs. of butter-fat taken out, and with the butter-fat 12 lbs. of milk, how many pounds of skimmed milk would be left? (Note. 4 lbs. of butter-fat in 12 lbs. of milk would make 16 lbs. of cream testing 25% fat, which is about the average for cream.)
2. If the 4 lbs. of butter-fat taken from the 100 lbs. of milk were sold for 33c. per pound, and the skimmed milk were worth 15c. per hundred, what would be the income from the 100 lbs. of milk?
3. If it costs 10c. per 100 lbs. more to market whole milk than to separate and market cream, at what price per 100 lbs. must 4% milk be sold to be as profitable as cream at 33c. per pound for butter-fat?

Out in the Fields with God.

The little cares that fretted me,
   I lost them yesterday
Among the fields, above the sea,
   Among the winds at play,
Among the lowing of the herds,
   Among the rustling of the trees,
Among the singing of the birds
   The humming of the bees.

The foolish fears of what may happen,
   I cast them all away
Among the clover-scented grass,
   Among the new-mown hay,
Among the rustling of the corn,
   Where the drowsy poppies nod,
Where ill thoughts die and good are born,
   Out in the fields with God.

—Anonymous.

APPENDIX

Average Composition of Some Common Feeds*
Amount in 100 Pounds

Feeds	Water	Ash	Crude protein	Carbohydrates Fiber	Carbohydrates Nitrogen free ext.	Fat
Alfalfa hay	8.6	8.6	14.9	28.3	37.3	2.3
Barley (grain)	9.3	2.7	11.5	4.6	69.8	2.1
Brewers' grains (dried)	7.5	3.5	26.5	14.6	41.0	6.9
Clover hay, alsike	12.3	8.3	12.8	25.7	38.4	2.5
Clover hay, crimson	10.6	8.8	14.1	27.3	36.9	2.3
Clover hay, Japan	11.8	5.8	12.1	25.9	41.6	2.8
Clover hay, red	12.9	7.1	12.8	25.5	38.7	3.1
Clover hay, sweet, (white)	8.6	7.2	14.5	27.4	40.1	2.2
Clover and timothy, mixed	12.2	6.1	8.6	29.9	40.8	2.4
Corn, dent	10.5	1.5	10.1	2.0	70.9	5.0
Corn, flint	12.2	1.5	10.4	1.5	69.4	5.0
Corn meal	11.3	1.3	9.3	2.3	72.0	3.8
Corn silage (well matured)	73.7	1.7	2.1	6.3	15.4	0.8
Corn stover (ears removed, very dry)	9.4	5.8	5.9	30.7	46.6	1.6
Cottonseed meal (good)	7.9	6.4	37.6	11.5	28.4	8.2
Cowpeas	11.6	3.4	23.6	4.1	55.8	1.5
Cowpeas hay	9.7	11.9	19.3	22.5	34.0	2.6
Germ oil meal (high grade)	8.9	2.7	22.6	9.0	46.0	10.8
Gluten feed (high grade)	8.7	2.1	25.4	7.1	52.9	3.8
Gluten meal	9.1	1.1	35.5	2.1	47.5	4.7
Johnson grass hay	10.1	7.5	6.6	30.2	43.5	2.1
Mangels	90.6	1.0	1.4	0.8	6.1	0.1
Milk, cow's (whole)	86.4	0.7	3.5	....	5.0	4.4
Milk, skim	90.1	0.7	3.8	....	5.2	0.2
Millet hay (common)	14.3	6.3	8.3	24.0	44.3	2.8
Mixed grasses, hay	12.8	5.6	7.6	28.8	42.7	2.5
Oats	9.2	3.5	12.4	10.9	59.6	4.4
Oats, straw	11.5	5.4	3.6	36.3	40.8	2.4
Oil meal (old process)	9.1	5.4	33.9	8.4	35.7	7.5
Potatoes	78.8	1.1	2.2	0.4	17.4	0.1
Rice meal	9.5	9.1	11.8	9.3	48.7	11.6
Rye	9.4	2.0	11.8	1.8	73.2	1.8
Soybeans	9.9	5.3	36.5	4.3	26.5	17.5
Soybeans hay	8.6	8.6	16.0	24.9	39.1	2.8
Sugar beets (roots)	86.3	1.1	1.6	1.0	12.6	0.1
Tankage (high grade)	7.5	12.0	60.5	4.2	2.7	13.0
Timothy hay	11.6	4.9	6.2	29.9	45.0	2.5
Wheat	10.2	1.9	12.4	2.2	71.2	2.1
Wheat bran	10.1	6.3	16.0	9.5	53.7	4.4
Wheat middlings (shorts)	10.4	4.4	17.3	6.0	57.0	4.9
Wheat screenings	10.2	3.9	13.3	7.4	61.1	4.1
Whey	92.5	1.6	1.0	....	4.6	0.3

*Compiled from Feeds and Feeding, Henry and Morrison, 15th Edition, 1915.

APPENDIX

Planting Information for Vegetables

PLANT	Soil	Depth	Plants Apart	Rows Apart	Days to Germinate	Days to Mature
Cabbage	Rich loam	3"	2'	3'		100-110
Cauliflower	Rich loam	3"	2'	3½'		105-115
Celery	Light rich	3"	8"	15"		84
Onion sets	Rich	3"	3"			
Tomatoes	Rich sandy loam	6"	3'	3'		90-100

SEEDS			Seeds to Foot			
Beans	Loamy	2"	4 to hill	24-30"	5-10	43-60
Beets	Sandy loam	1"	10	18"	7-10	60-70
Carrots	Light	½"	6-12	15"	12-18	95-100
Celery	Light rich	¼"	12	15"	10-20	160-175
Corn	Rich loam	1½"	5 to hill	3'	5-8	55-75
Cucumbers	Rich	1"	4 to hill	6'	6-10	50-75
Lettuce	Rich loam	¾"	15	15"	6-8	60-65
Onion seeds	Good	¼"	15	15"	7-10	100-110
Parsley	Rich loam	¼"	6	18"	20-30	98
Parsnips	Light	½"	15	15"	10-20	125-140
Peas	Light	2½"	8	3'	6-10	60-80
Radishes	Fertile sandy clay	¾"	20	15"	3-6	21-35
Salsify	Light	1"	8	18"	7-12	130-140
Spinach	Rich sandy loam	1"	12	15"	6-10	21-30
Turnips	Open soil	¼"	15	15"	4-8	60

Distances Apart for Planting Fruits

Trees	Distance	Bush or Vine	Distance
Apples	30-40 ft.	Blackberries	4½-7 ft.
Apricots	15-20 ft.	Cranberries	1-2 ft.
Cherries	15-25 ft.	Currants	4-4½ ft.
Oranges	25-30 ft.	Gooseberries	4-4½ ft.
Peaches	15-20 ft.	Grapes	6-12 ft.
Pears	20-30 ft.	Raspberries, black	3½-5 ft.
Plums	15-20 ft.	Raspberries, red	3½-4 ft.
Quinces	10-12 ft.	Strawberries	1½-3 ft.

Haecker's Feeding Standard for the Dairy Cow

	Daily allowance of digestible nutrients		
	Crude Protein	Carbohydrates	Fat
	Lbs.	Lbs.	Lbs.
For support of the 1,000-lb. cow	0.7	7.0	0.1
To the allowance for support add:			
For each lb. of 3.0 per cent milk	0.040	0.19	0.015
For each lb. of 3.5 per cent milk	0.042	0.21	0.016
For each lb. of 4.0 per cent milk	0.047	0.23	0.018
For each lb. of 4.5 per cent milk	0.049	0.26	0.020
For each lb. of 5.0 per cent milk	0.051	0.27	0.021
For each lb. of 5.5 per cent milk	0.054	0.29	0.022
For each lb. of 6.0 per cent milk	0.057	0.31	0.024

Quantities of Seed Required to the Acre

Name	Quantity of Seed	Name	Quantity of Seed
Alfalfa	6-12 lbs.	Orchard grass	20-30 lbs.
Alsike	8-20 lbs.	Oats	2-4 bu.
Barley	1½-2½ bu.	Parsnips	6-8 lbs.
Beans	1-2 bu.	Peas	2½-3½ bu.
Blue grass	10-15 lbs.	Potatoes	5-10 bu.
Broom corn	1-1½ bu.	Rice	2-2½ lbs.
Buckwheat	¾-1⅓ bu.	Red Clover	10-16 lbs.
Carrots	4-5 lbs.	Rye	1-2 bu.
Corn	¼-1 bu.	Timothy	12-24 qts.
Flax	½-2 bu.	Turnips	2-3 lbs.
Hemp	1-1½ bu.	Wheat	1¼-2 bu.
Millet	1-1½ bu.	White Clover	3-4 lbs.

INDEX

Acid phosphate, 36
Accounts:
  With a cow, 329
  With a garden, 126
  With live stock, 326
Agricultural Engineering, 247
Agricultural papers, 307
Air:
  Needed in soil, 26
  Plant food in, 13
Alfalfa:
  Advantages of, 100
  Curing, 103
  Cutting, 103
  Feeding value of, 101
  Inoculation of, 102
  Pasturing, 101
  Seed, 101
  Soil for, 102
  Sowing, 102
Apple blight, 148
Apples:
  Adaptability of, 142
  Culture of, 143
  Mulching, 144
  Planting of trees, 143
  Pruning of trees, 144
  Soil for, 142
  Spraying, 149
Apple scab, 149
Arsenate of lead, 149
Ash, 164
Asparagus, 130

Babcock test, 204
Bacteria:
  Care against, 299
  In milk, 203
  On clover roots, 96
Balanced ration, 165
Barley:
  Culture of, 49
  Harvesting, 49
  Use of, 49

Barns, 261
Beans, 132
Bedding, 190
Bees, 244
Birds, 243
Blue grass, 107
Boll weevil, 153
Bookkeeping, 324
Bordeaux mixture, 149
Boys' and Girls' Club, 272
Breeds:
  Of cattle, 182
  Of horses, 167
  Of poultry, 234
  Of sheep, 206
  Of swine, 218
Brome grass, 106
Brood sow, 222
Buildings:
  Convenience of, 262
  Cost of, 260
  Importance of, 260
  Maintenance of, 262
  Planning of, 261
  Ventilation of, 160, 171, 300
Bull thistle, 122
Bulletins, 124, 244, 246
Burdock, 122
Butter:
  Making, 331
  Marketing, 283
Butter-fat, 205

Cabbage worm, 152
Calcium:
  Use of in plants, 34
  Supply of, 37
Canada thistle, 124
Carbohydrates, 164
Carbon dioxide, 13
Carrots, 86
Cattle:
  Breeds of, 182
  Care and management of, 189

Cattle: Cont'd
  Classes of, 182
  Disease of, 189
  Feeding of, 191
  Pure-bred, 182
  Shelter for, 189
  Types of, 182
Celery, 132
Chickens, 232, 299
Chinch bugs, 152
Chlorophyll, 13
Chores, 156
Churning, 332
Clay, 11
Clover:
  Adds nitrogen to soil, 96
  Alsike, 93
  Curing, 98
  Getting a catch of, 94, 321
  In rotation, 321
  Mammoth, 92
  Medium red, 92
  Roots and bacteria, 95
  Varieties of, 92
  White, 93
Clubs:
  Boys' and Girls' 273
  Farmers', 276
Cockle, 118
Codling moth, 152
Comfort of animals, 225
Community activities, 272
Concentrates, 164
Co-operation, 279, 281, 334
Corn:
  Climate for, 53, 69
  Culture of, 61, 66
  Depth to cultivate, 66
  Embryo of, 57
  For silage, 78
  Grading seed, 62
  Importance, 52
  In rotation, 321, 324
  Parts of kernels of, 56
  Planting, 63
  Rag doll tester, 60
  Selection of seed, 68, 71
  Shape of ear, 72
  Size of ear, 72
  Size of kernels, 54

Corn: Cont'd
  Smut, 148
  Storing, 73
  Testing seed, 58, 59
  Varieties of, 70
  Yield of, 52
Corn hangers, 77
Corrosive sublimate, 150
Cot for hogs, 224
Cotton:
  Diseases of, 151
  Picking and ginning, 112
  Planting and culture, 112
  Soil for, 112
County agents, 290
Cowpea, 109
Cows:
  Account with, 329
  Culling, 206
  Nutrient requirements for, 191
  Rations for, 193
  Succulent feed for, 197
  Testing of, 206
  Types of, 185 *et seq.*
  Water for, 189
Cream, 205, 333
Creamery, 283
Creep, 224
Cucumbers, 133
Cultivated crops, 52, 312, 314
Cultivation:
  Depth of for corn, 66
  Of apple trees, 143
  Of corn, 61
  Of fiber crops, 112
  Of potatoes, 85
  Of raspberries, 141
  Of strawberries, 138
  Reasons for, 63
  Results of deep, 66
Cultivator, 67
Curing hay, 98
Curled dock, 124

Dairy products, 283
Diseases:
  Of cotton, 151
  Of grains, 146
  Of potatoes, 150
Disking, 28

INDEX

Drainage:
  Of fields, 254
  Of home grounds, 298
  Of roads, 250
  Surface, 255
  Tile, 255
Drill, 29
Dry farming, 24
Drying seed corn, 75

Educational advantages, 277
Eggs, 286
Ensilage, 198
Eradication of weeds, 122
Ewes, 213, 216
Exercise for stock, 190

Farmer, standing of, 306
Farm home, 293
Farm management, 306
Farmstead:
  Arrangement of, 303
  Location of, 303
  Size of, 303
Fat, 164
Feed:
  Change of, 178
  Comparison of, 166
  Composition of, 164, 199
  Fall, 23, 227
  Kinds of, 164
  Requirements, 163
  Selection of, 163
  Source of, 163
  Summer, 230
  Value of, 328, 330
Fencing:
  Building of, 268
  Cost of, 270, 316, 319
  For sheep, 214
  For swine, 225
  Investment in, 267
  Kinds of, 265
  Posts for, 266
Fertilizers:
  Animal manure, 37
  Amounts produced, 38
  Complete, 37
  Composition of, 38
  Need of, 31

Fertilizers: Cont'd
  Plant food in, 31
  Uses of, 34
Fiber crops, 112
Field crops, 310
Field peas, 109, 227
Fields, 315, 317
Flax:
  Harvesting, 113
  Planting, 113
  Soil for, 113
Flax wilt, 148
Flies, 190, 299
Formaldehyde, 147
Forage crops, 108
Garden:
  Account with, 126
  Income from, 125
  Plan and preparation of, 127
  Value of, 125
Germination, 44, 58
Grain crops, 310, 322
Grass crops, 89, 310
Grasshoppers, 152
Gravel, 11

Hay and pasture crops
  Advantages of, 89
  Cleaning crop, 120
  Cock covers for, 99, 104
  Cost, 89
  Curing, 98
  Cutting, 97, 105
  Importance of, 89
Harrowing, 28
Harvesting:
  Barley, 49
  Flax, 113
  Oats, 49
  Root crops, 87
  Sugar cane, 111
  Wheat, 43
Healthfulness of the home, 298
Heat in the soil, 27
Hemp:
  Culture of, 113
  Soil for, 113
Hog cholera, 221
Hogging off crops, 226, 228
Hogs. (See swine)

Home, the, 293
Honey, 245
Horses:
   Breeds of, 167
   Care and management of, 171
   Cost of labor of, 172
   Feeding of, 174 *et seq.*
   Shoeing, 172
   Types of, 168
Horse-radish, 130
Humus, 10
Hydrogen, 13

Income of American farmers, 307
Insect pests, 146
Insects:
   Classes of, 152
   Control of, 152
   Destroyed by plowing, 20
   Habits of, 152
   Remedies for, 152
Interest, 327
Irrigation, 256

Johnson grass, 107

Kinghead, 118

Labor:
   General, 328
   In dairying, 330
   With sheep, 231
   With swine, 226
Land rent, 325
Lawns, 294
Legumes, 96
Lettuce, 131
Light, 160
Lime, 34, 37
Lime-sulphur mixture, 147
Live stock:
   Care and management of, 156
   Classes of, 155
   Importance of, 155
   In relation to soil, 155
   Shelter for, 160
Loam, 11

Machinery:
   Investment in, 258
   Shelter for, 259
   Use of, 259

Mangels, 86
Manure:
   Amounts produced, 38
   Composition of, 38
Manure spreader, 82
Marketing, 284, 331
Meadow. 323
Melons, 133
Milk:
   As human food, 202
   Care of, 203
   Composition of, 202
   Sampling, 207
   Selling, 333
   Testing, 204
   Weighing, 206
Millet, 108
Miscellaneous crops, 108
Moisture, 20, 63
Morning-glory, 124
Mulch, 64, 144
Mules, 156

Nests, 240
Nitrogen:
   Added to soil, 96
   Need of in plants, 32
   Sources of, 33
   Nutrients, 65, 191
Nodules. 96

Oats:
   Harvesting, 49
   Importance of, 45
   In rotation, 321
   Smut of, 147
   Uses of, 45
Oils, 164
Onions, 130
Orchard, 142
Orchard grass, 107
Outside feeding, 159
Overrun, 332
Oxygen, 13

Parcel post, 307
Paris green, 153
Parsnips, 130
Pasteurizing milk, 203

INDEX

Pasture:
  Alfalfa, 101
  Benefit to soil, 323
  Blue grass, 231
  Brome grass, 231
  Field peas, 231
  Rape, 231
  Red clover, 231
  Rye, 231
  Timothy, 105
  White clover, 231
Peas, 129, 131
Phosphate rock, 36
Phosphorus:
  Need of in plants, 33
  Sources of, 35
Pigeon grass, 120
Pigweed, 121
Planning:
  Buildings, 261
  Farms, 315
Plant food:
  Amount of, 15
  Available, 15
  In fertilizers, 31
  In the air, 13
  In the soil, 13
  Liberated by plowing, 21
  Soluble, 14, 16
Plant diseases:
  Loss from, 146
  Prevalence of, 146
  Remedies, for, 146 *et seq.*
Planting:
  Depth to plant, 29
  Time to plant, 28
  Plant structure, 39
Plant lice, 152
Plowing:
  Condition of soil for, 21
  Deep, 24
  Fall, 22
  Objects of, 19
  Time for, 21
Plum curculio, 152
Pork production, 220
Posts, 266
Potassium:
  Need of in plants, 34
  Sources of, 36

Potatoes:
  Cultivating, 83
  Cutting seed, 83
  Diseases of, 82
  Importance of, 80
  Planting, 83
  Seed, 81
  Soil for, 83
  Spraying, 85
  Sprouting, 81
Potato blight, 151
Potato bugs, 152
Potato scab, 150
Potato wilt, 150
Poultry:
  Breeds of, 234
  Care of, 236
  Feeding, 241
  Houses, 238
  Importance of, 232
Price of land, 306
Prizes, 274
Protein, 164
Pruning:
  Apple trees, 144
  Raspberry bushes, 141

Quack grass, 121

Radishes, 131
Rag doll tester, 60
Ragweed, 118
Rape, 108, 231
Raspberries:
  Adaptability of, 139
  Cultivation of, 141
  Planting, 140
  Propagation of, 139
  Pruning of, 141
  Soil for, 139
  Varieties of, 139
  Winter protection of, 141
Ration:
  A good, 196
  A poor, 195
  Feeding a, 195
  To compound a, 193
Rations. (See feeds, etc.)
Red top, 107
Rent, 325

Rhizoctonia, 150
Rhubarb, 130
Rice, 110
Roads:
    Construction of, 249
    Cost of, 247
    Earth, 250
    Gravel, 251
    Maintenance of, 252
    Sandy, 254
    Use of, 247
Roosts, 239
Root crops:
    Culture of, 86
    Harvesting, 87
    Importance of, 86
Roots, 199
Rotation of crops, 91, 153, 308, 312, 317, 320, 322
Roughage, 164, 175, 193
Rural mail delivery, 307
Rust, 29, 146
Rutabagas, 86
Rye:
    Culture of, 50
    Importance of, 50
Rye grass, 107

Salt, 189
Sand, 11
Sanitation:
    In the care of milk, 203
    In the home, 298
Scale insects, 152
School gardens, 289
Schools, 272
Seed:
    Grading, 44
    Importance of good, 40
    Parts of a, 41
    Pure, 43
    Selection of, 41
    Test of good, 41
    To remove weed, 43
Seed bed, 26
Seed corn:
    Selection of, 68
    Storing, 73
    Testing, 58
Septic tank, 307

Shade, 189
Shade trees, 295
Sheep:
    Care and management of, 213
    Feeding of, 215
    Fencing for, 214
    For fattening, 215
    Shelter for, 213
    Types of, 206
Shelter:
    Cost of, 329
    For live stock, 160, 189, 214, 223, 238
    For machinery, 259
Shocking and stacking, 44
Silage:
    Corn, 78
    Cutting, 80
Silo:
    Advantages of, 263
    Cost of, 264
    Importance of, 263
    Kinds of, 264
    Size of, 265
Slatted shelves, 77
Smut:
    Corn, 148
    Covered of wheat, 146
    Grain, 146
    Loose of barley, 146
    Loose of oats, 146
    Loose of wheat, 146
Social advantages, 276
Soil moisture, 14
Soils:
    Classes of, 10
    Clay, 12
    Loam, 11
    Origin of, 9
    Parts of, 9
    Sandy, 12
Soy bean, 109
Sow, brood, 222
Sow thistle, 124
Split-log drag, 252
Spraying:
    Fruits, 152
    Potatoes, 85
Squash, 133
Squash bugs, 152

INDEX

Starch, 164
Sterilizing milk, 202
Storing seed corn, 73
Straw, 36
Strawberries:
    Adaptability of, 136
    Culture of, 138
    Planting, 138
    Soil for, 136
    Varieties of, 136
Subsurface packing, 25
Succulent feed, 197
Succulent food, 134
Sugar, 164
Sugar beets, 86
Sugar cane, 111
Surface mulch, 25, 64
Sweet corn, 132
Swine:
    Care and management of, 220
    Feeding of, 226
    Fencing for, 225
    Shelter for, 223
    Types of, 218
System in work, 157

Teachers, 272
Telephone, 307
Testing cows, 206
Testing milk, 204
Threshing, 44
Tile, 255
Tillage:
    Objects of, 19
    Time for, 21
Timothy:
    Cutting, 105
    Feed value of, 105
Tobacco extracts, 153
Tomatoes, 133
Tuberculosis, 190
Turnips, 86

Vegetable matter:
    Adding to soil, 17
    Decay of, 17

Vegetables:
    Varieties of, 129
    Marketing of, 130
Ventilation, 160, 171, 300
Vetch, 110

Waste land, 305, 316
Water for cattle, 159, 189
Wax, 164
Weaning pigs, 224
Weeds:
    Classes of, 121
    Destroyed by plowing, 20, 64
    Habits of, 121
    Mounting, 116
    Seeds, 46, 117, 119
    Specimens of, 116
    To eradicate annual, 122
    To eradicate biennial, 122
    To eradicate perennial, 124
Weed seeds, 43, 117, 119
Wheat:
    Harvesting, 46
    Importance of, 45
    Kinds of, 46
    Place of, 45
    Shocking and stacking, 47
    Soil for, 46
    Sowing, 46
    Threshing, 47
Wild buckwheat, 119
Wild oats, 118
Wild pea, 119
Windbreaks:
    Planning of, 296
    Trees for, 296
    Value of 296
Wool, 210

Yield:
    Dependent on seed bed, 26
    Of corn, 52
    Of hay, 89
    Of oats, 48
    Of wheat, 45

POPULAR FRUIT GROWING

Samuel B. Green, Late Professor of Horticulture, University of Minnesota, 4th edition revised by Le Roy Cady, Associate Professor of Horticulture, University of Minnesota. Cloth, 12 mo., 300 pages, illustrated..$1.00
Paper... .50

All phases of fruit growing are discussed in a clear and interesting manner.

The book deals with every branch of the work, from the preparation of the soil to marketing, and explains every operation in detail. The principal topics treated are Factors in Successful Fruit Growing, Orchard Protection, Insects Injurious to Fruits, Diseases, Harvesting and Marketing, Principles of Plant Growth, Propagation, Pome Fruits, Stone Fruits, Grapes, Small Fruits and Nuts. An appendix contains many formulas and other matters of importance.

This text has fully justified itself by years of popularity. It has maintained its hold in many schools of high standing and over a wide area solely on account of its perfect adaptability to the very uses for which it was intended.

VEGETABLE GARDENING

Samuel B. Green, Late Professor of Horticulture, University of Minnesota, 12th edition revised by Le Roy Cady, University of Minnesota. Cloth, 12 mo., 336 pages, illustrated...................$1.00
Paper... .50

Vegetable Gardening presents the subject of gardening in a clear and definite way. Its aim is to enable the student to understand the fundamental principles involved in each practical undertaking and to recognize the importance of careful, systematic and scientific methods in the growing of vegetable crops. The entire plan of the book is noteworthy. Each garden crop is discussed fully from seed to marketing. There are also valuable treatments of Gardening in General, the Home Garden, Fertilizers, Tillage, Seeds, Transplanting, Development of Varieties, Glass Structures Insects and Insecticides, Marketing, Exhibiting, etc.

The above features make this a textbook especially adapted to agricultural high schools and other secondary agricultural schools and for beginning classes in colleges. For years it has been regarded as a leading authority.

PROBLEMS IN CARPENTRY

Louis M. Roehl, Director of Farm Mechanics, Milwaukee County School of Agriculture and Domestic Science, Wauwatosa, Wis. Cloth, 7¼x8½ inches, 112 pages......................................$1.00

A Course in Practical Carpentry for Manual Training Classes in Public Schools. It comprises a series of sixteen projects fully illustrated and covering all the features of house constructing, including practice in framing, flooring, lathing, plastering, calcimining, painting, varnishing, etc.

Full scale lumber is used for each project. Actual building conditions are made the basis of the work. The course is exceptionally practical and exactly adapted to industrial classes.

A plan for each project is given along with two or three different views of the finished project.

RURAL EDUCATION

A. E. Pickard, Superintendent of Associated Schools, Cokato, Minnesota. Cloth, 12 mo., 432 pages, illustrated............$1.00

This text comprehends a complete course of study for modern rural schools. With introductory chapters on the training of teachers and school management, the book sets forth fully four daily programs for the reorganized method of conducting such schools. The academic and industrial work for every division and every subject are explained in detail. The chapters on industrial work are profusely illustrated. The whole course in sewing is outlined and illustrated; fifty recipes for lunches are given; and twenty-five projects for manual training. Numerous outlines for agricultural booklets are shown, and the scheme for association and consolidation of schools is fully described.

This work will prove intensely interesting to everyone whose vocation concerns the rural schools in any way.

Bound separately are furnished the chapters on the different phases of industrial work, and those including the booklet work.

DOMESTIC SCIENCE: PRINCIPLES AND APPLICATION

Pearl L. Bailey, Supervisor of Domestic Science, St. Paul (Minnesota) Public Schools. Waterproof cloth, 12 mo., 357 pages, illustrated..$1.00

A textbook for students of Domestic Science so arranged and graded as to cover the work of two successive years of standard courses as established in the average public school. The aim has been to present a course of systematic cooking reinforced with an explanation of the scientific principles involved. The attention of the student is not distracted by a multiplicity of irrelevant subjects; but the book contains many suggestions as to equipment, all sorts of school and business lunches, and serving. There are also score cards for contests, tables of comparative food values and substitutes, lists of poisons and their antidotes, helpful hints for aid to the injured, etc.

The recipes are abundant, simple and economical; and the illustrations are actual photographs of classroom work and conditions.

The most popular book of its kind ever published.

ELEMENTS OF FARM PRACTICE

A. D. Wilson, Director of Agricultural Extension, University of Minnesota, and E. W. Wilson. Cloth, 12 mo., 352 pages, Illustrated..$1.00

This volume is a thorough revision of "Agriculture for Young Folks." The book, while maintaining the excellent and characteristic features of the former edition, has been largely rewritten and brought thoroughly up to date. Many new topics have been introduced and much care has been taken to preserve, according to its relative importance, a proper proportion of subject matter. The book is profusely illustrated with new plates, printed from large new type and handsomely bound.

The plan of the authors is to emphasize matters of importance as actually practiced in the laboratory of the farm and the home. Along with its adaptability to conditions it is delightfully teachable.

There is no simpler, more attractive, interesting or practical book upon the market for use in rural or graded schools. It is also distinctively the book for use in normal schools.

AGRICULTURAL ENGINEERING

J. B. Davidson, Professor of Agricultural Engineering, Iowa State College, and joint author of Farm Machinery and Farm Motors. Cloth 12 mo., 548 pages, 342 illustrations....................$1.50

This volume has been written primarily as a textbook for secondary schools of agriculture and for colleges where only a general course can be offered. About one half of the text is devoted to farm machinery and motors, while the other half comprises chapters on Surveying, Drainage, Irrigation, Roads, Farm Structures, Farm Sanitation and Rope Work. The subjects treated do not involve a technical knowledge or equipment beyond the attainments or facilities of the ordinary high school student. The language is simple and the principles easily comprehensible. This book will furnish one of the most practical courses that can be offered. Successful farm management is largely dependent upon these engineering facts and principles.

Professor Davidson's wide experience as an investigator and teacher has placed him foremost among agricultural engineers.

FIELD MANAGEMENT AND CROP ROTATION

Edward C. Parker, Formerly Assistant Agriculturist, Minnesota Agricultural Experiment Station; Special Agent, Bureau of Statistics, U. S. Department of Agriculture; and Agricultural Expert with the Government of Manchuria. Cloth, 12 mo., 512 pages, illustrated..$1.50

This book meets a growing need for a definite course in the handling and management of fields. It deals with Planning and Reorganizing Fields, Crop Rotation Systems, Soil Amendment with Fertilizers, Relation to Animal Husbandry to Soil Productivity, but does not cover the detailed financial matters usually included in Farm Management texts. Besides much Experimental Evidence, there are chapters on Plowing Practice, Soil Inoculation, Seed Selection, Improved Crop Varieties, Fungus Diseases, Weed Eradication, together with an Appendix full of very valuable material for reference.

The text will fill an important place in secondary agricultural courses.

CHEMISTRY OF THE FARM AND HOME

W. E. Tottingham, Assistant Professor of Soils, University of Wisconsin, and J. W. Ince, Assistant Professor of Soils, North Dakota Experiment Station. Cloth, 12 mo., 450 pages, illustrated....$1.25

There has been a demand for a text that would furnish sufficient general theory to satisfy the demands of college entrance requirements and yet that would receive its illustration and application from practical affairs of the farm and the home. It is the purpose of this volume to meet this need. Besides a general introductory chapter, the subjects treated are: Water and Its Constituent Elements, the Atmosphere and Its Chief Constituent Nitrogen, Other Important Nonmetals, Important Metals, the Plant and Its Products, The Soil, Fertilizers, Farm Manure, The Animal and Its Products, Feeding of Animals, Dairy Products, Human Food and Dietetics, and Miscellaneous Materials of Importance in Daily Life.

Summaries and questions are given at the close of each chapter and abundant experiments for laboratory practice.

Combine general and industrial chemistry in this book.

FIELD CROPS

A. D. Wilson, Superintendent of Farmer's Institutes and Agricultural Extension, University of Minnesota, and **C. W. Warburton,** Agronomist, Bureau of Plant Industry, U. S. Department of Agriculture. Cloth, 12 mo., 544 pages, illustrated $1.50

This book embodies a thorough discussion of Grain Crops, Forage Crops, Roots, Tubers and other miscellaneous crops, Plant Growth, Rotation, Weeds, etc., accompanied by a treatment of History and Classification, Importance, Production, Harvesting, Marketing, Insects and Diseases, Relation to Other Crops, Utilization, Improvement, Judging, Supplementary Reading and Laboratory Exercises.

The aim has been to make a book that is intesely practical.

The authors are men of natural reputation who have given the result of long experience and much practical experiment.

The book is neither elementary nor advanced and technical. The style is simple, clear and concise. It is the most usable and popular text that has been written on this subject.

BEGINNINGS IN ANIMAL HUSBANDRY

Charles S. Plumb, Professor of Animal Husbandry, Ohio State University, Author of Types and Breeds of Farm Animals, Indian Corn Culture, etc. Cloth, 12 mo., 393 pages, illustrated....$1.25

A full discussion of Horses, Cattle, Sheep, Swine and Poultry, presenting the essential consideration in Types, Breeding, Judging, Care and Management and including fitting chapters on Pedigrees, Composition of Plants and Animals, Value of Feeds, Feeding Standards, and Live Stock Equipment. The book is so systematically arranged, splendidly developed and admirably illustrated that it appeals strongly and greatly facilitates classroom instruction.

Professor Plumb has long been known as one of the leading instructors and investigators in the sphere of animal husbandry, as a judge of live stock and as a reliable and interesting writer on subjects covered by his study. In this new volume he has concentrated the result of his many years of teaching and experimentation as well as much valuable information gleaned from other sources.

SOILS AND SOIL FERTILITY

A. R. Whitson, Professor of Soils, University of Wisconsin, and **H. L. Walster,** Instructor in Soils, University of Wisconsin. Cloth, 12 mo., 320 pages, illustrated................................$1.25

This volume teaches how to understand soils and how to analyze and handle them in a practical way. The authors have presented one of the most systematic, careful and complete discussions of soils and soil fertility ever published. Their wide laboratory and classroom experience and especially their extensive and close study of practical field problems make them eminently qualified as worthy authorities and insure a highly useful volume.

The text is intended as an introduction to the study of soils. It does not go into unnecessary scientific detail of structure and chemical composition, but discusses in a simple, practical way the best methods of handling the land to produce the best crops. Its principles are immediately applicable to farming.

There are copious laboratory exercises.

www.ingramcontent.com/pod-product-compliance
Lightning Source LLC
Chambersburg PA
CBHW062318220526

45469CB00008B/2548